CONTENTS

ACKNOWLEDGEMENTS

We thank Gill Greensides, Barbara Wilkinson and Michele Stone for their care and skill in producing the typescript for this book.

MG & PJG

SCIENCE, TECHNOLOGY
AND SOCIETY TODAY

For F.R. Jevons

SCIENCE, TECHNOLOGY AND SOCIETY TODAY

Editors
Michael Gibbons and Philip Gummett

MANCHESTER UNIVERSITY PRESS

Copyright © Manchester University Press 1984

Published by
Manchester University Press
Oxford Road, Manchester M13 9PL, UK
and 51 Washington Street, Dover
New Hampshire O3820, USA

British Library cataloguing in publication data
Science, technology and society today.
   1. Science – Social aspects
   2. Technology – Social aspects
   I. Gibbons, Michael   II. Gummett, Philip
   306'.54   Q175.5

ISBN 0-7190-0905-7

Library of Congress cataloging in publication data
Main entry under title:
Science, technology, and society today.
   Includes bibliographies.
   1. Science – Social aspects. 2. Technology – Social
aspects. I. Gibbons, Michael, 1939–   II. Gummett,
Philip.
Q175.5.S3735 1983   303.4'83   83–20639
ISBN 0-7190-0878-6 (pbk.)

Printed in Hong Kong
by Wing King Tong Co. Ltd.

INTRODUCTION

Michael Gibbons & Philip Gummett

Technological unemployment, genetic engineering and the prospect of a nuclear holocaust are three concerns of the present age in which science and technology are implicated in some degree. They are but examples of a much larger set of contemporary questions the understanding of which requires examination of the social, political and economic interrelations of science and technology. Such examination in turn demands that we reflect upon what we mean when we talk about 'science' and 'technology', that we consider how these fields of human activity develop, and that we reach some understanding of what is meant by the claim that the societies of Western civilization became, at a certain point in their history, 'scientific'. The historian Herbert Butterfield, in his book *The Origins of Modern Science*, concluded that the scientific revolution of the sixteenth and seventeenth centuries outshone 'anything since the rise of Christianity and reduce[d] the Renaissance and the Reformation to the rank of mere episodes, mere internal displacements within the system of medieval Christendom'. Such a claim implies that science has brought about a fundamental transformation of civilization.

The field of Science, Technology and Society (STS) is not alone in being concerned with the exploration of the nature of this transformation; so are many other established academic disciplines — for example, history, philosophy and political theory. What is different about STS is that it is concerned to explore not so much the *impact* of science *on* society as the *interrelationship* of scientific activities *with* other social, economic and political activities. And, because so many current events seem to be connected with scientific activities, many who work in STS have tended to focus on the *contemporary* interrelationship between science and society. They may have focused on *opportunities* apparently created by science, such as the applications of scientific theories or ideas to manufacturing industry, to the delivery of health care, to national defence or to the search for alternative sources of energy. Alternatively, they may have focused on *problems*, like the control of environmental pollution, the carcinogenic properties of new drugs, or the unemployment effects of new technologies like microprocessors and biotechnologies. In brief, the opportunities and problems presented by contemporary science are seen as part of a larger process in which there is a two-way flow of effects between science and society. As a corollary, it is necessary to study not only the way science affects society but also the possibility of social influences upon the rate and direction of scientific activity.

As an example, consider the writings of scientists and others after the bombing of Hiroshima and Nagasaki. In the record of the time there is a clear recognition by many, particularly scientists, that the Bomb marked a transition to a new relationship between science and politics. Indeed, doubts began to grow about the implications of the further development of science and its associated technological applications. These doubts derived in part from pangs of conscience over what scientists themselves had made possible. But they were also expressions of concern over the fact that the generation and application of scientific ideas now depended so heavily on governmental, especially military, support, backed up by industrial muscle.

The significance of this transformation was easily overlooked, partly because the enormous military research and development programmes of many nations were conducted out of sight of the public, and partly because it was obscured by the boost given to science in the immediate post-war years as nations turned to science to help rebuild or expand their economies. Science, having 'won the war', was now going to help 'win the peace', and it was easy to lose sight of the implications of the first atomic explosions as an economic boom began, based upon technologies developed during the war – nuclear power, jet aircraft, plastics, electronics – and their attendant sciences, chemistry and physics. Given widespread belief that the creation of new industries and the refurbishment of the old depended upon the application of science, anyone who questioned the overall pattern and purpose of scientific developments went unheeded by those who considered that scientists only make discoveries, and that their application, for good or ill, is the responsibility of others.

This attitude of mind – that scientists *discover* but others *apply*, and hence that the principal flow of effects is from science *to* society – is still widespread. Indeed, its roots may be traced back to the origins of modern science itself. The campaign of cultural propaganda attempted by Galileo to enlist the support of the church for the 'new science' had its, albeit less spectacular, counterparts in other countries. Science, it must be remembered, did not emerge fully formed; neither was it immediately accepted by everyone as an unequivocal good. It first had to achieve a measure of social acceptance for its activities and this involved making clear what it was and was not about, and in what ways scientific knowledge differed from traditional sources of knowledge. One way of doing this was to insist (it not always being possible to demonstrate) that the ideas and theories of the new science were derived methodically by induction from the facts of experience. This 'Baconian' approach sought to free science from intellectual and social dependence upon established authorities who, too often, were metaphysicians or theologians. By arguing that their ideas had their roots in the empirical world only, scientists (then called natural philosophers) were able to distance themselves from philosophical and religious controversy, thereby winning a measure of political and social independence through which to pursue their activities. Thus, the earlier strategy of *involvement* with society followed by Galileo, and to some extent by Bacon, brought forth the need for a counter-strategy — one of independence from society.

To assist them in adopting a strategy of independence, scientists appropriated — probably via the philosophy of Descartes — the Aristotelian idea that science was the fruit of contemplation, characterized by a spirit of detached, disinterested inquiry into the empirical realm. The outcome of this activity also had some of the flavour of Plato's realm of pure ideas. Into this realm, social, political and economic concerns did not enter.

During the eighteenth and nineteenth centuries we can see the emergence of a tension that has yet to be resolved, concerning the attitude of scientists towards the usefulness of science. During this time, scientists were careful not to stress too much their relationships with industry or the military. They were seeking autonomy for their activities. On the other hand, to get social support there had to be some perception that the fruits of scientific activity could have useful results. One resolution of this dilemma was to assert that science contributed only at the discovery stage; others, industrialists for example, could apply the results. In this way, science had access to social utility without being seen to be too closely involved with industrial activity which, in England and the United States for example, would have affected its ability to develop in a university context. Few noted the obvious paradox of this position; that, if scientists were to be distanced from the 'evil' effects of the applications of scientific ideas, so too should they receive no credit for the 'good', or socially beneficial, effects of their activities.

The effect of all this was that throughout the eighteenth and nineteenth centuries scientists were preoccupied with building up the edifice of the scientific enterprise. During this time science was institutionalized in most of the countries of western Europe, a disciplinary structure with reasonable parallels to that of today was becoming visible and it was just becoming possible for a person to earn a living by practising science. Thus, by the twentieth century we can speak of a fully professionalized scientific community, specialized in terms of its subject matter but united in a common concern to promote strictly empirical methods of inquiry, and sharing a strong presumption that the results of their labours were in themselves ethically neutral.

The atomic explosions of 1945 ended all that decisively and dramatically. Since then, the claim of scientists to ethical neutrality has been harder to sustain. Moreover, the Manhattan Project prompted once again a more general examination of the relationship between science and society. There was no doubt that in the case of the atomic bomb, science had affected society, but it was equally clear that political and military demands had radically influenced the development of the relevant parts of science and technology. And the subsequent history of weapons development confirms the closeness and mutuality of the relationship between science and society. The development of thermonuclear weapons, for example, is a clear instance of science and scientists both guiding and being guided by political and military requirements and, to that extent, social objectives. Indeed, it is possible that historians may come to understand the development of nuclear physics in terms of the mutual interaction of scientists and the military, of science and society.

We have dwelt upon the issue of the Bomb because it illustrates clearly the *interdependence* of science and government. We can go on to ask whether this case is unique or whether it is an instance of a more general situation. Is it possible that science and society have always been mutually shaped through a complex series of interactions? The belief that this might be so, and that the hypothesis is worth testing, constitutes the cognitive ground of the field of STS.

The dimensions of STS

STS investigates a very complex phenomenon along many different dimensions, which we must now identify. The example of the Bomb concerns the interaction between science and government. There are many other areas where government has taken the lead in the development of science; radar and computation (electronics) or supersonic aircraft (aerodynamics) would be examples. Often, government support takes the form of establishing a national laboratory to pursue research of a given kind. But industry, too, through its major industrial research laboratories influences the development of science; witness the scientific efforts of Bell Laboratories in semi-conductor physics, Imperial Chemical Industries (ICI) in both organic and inorganic chemistry or CIBA-Geigy in biology and biochemistry. The rate and direction of scientific discovery in physics, chemistry and biology cannot properly be understood independently of the endeavours of large corporations such as these, and hence can be seen to have intimate connections with economic forces.

The political and economic modes of interaction between science and society take place today in the context of a fully professionalized scientific community. Indeed, it is one of the novelties of the scientific enterprise in the twentieth century that the community is composed of highly trained people who earn their living by working full time at a scientific career. The gifted amateurs supported largely by their own resources and working more or less independently have in large measure been replaced by professionals who work in teams in laboratories supported entirely by industry or government. A considerable part of the research activities of the STS community is given over to unravelling the origins of the various specialist groups which make up the scientific community at any one time. The field of social history of science, in particular, seeks to explain the development of science in terms of its social institutions and its social structure. Indeed, some social historians of science seek to account for the emergence of specific disciplines solely in terms of the dynamics of social institutions.

The social history of science has, in the last twenty years, been trying to effect a revolution in how we should think about the nature of scientific knowledge. As a result, the view of science as a purely intellectual activity producing value-free knowledge and universal truths is being challenged by alternative views which locate the impetus of scientific development in social, economic or political institutions. It is only recently that attempts have been made to trace the identification of problems and the formulation of hypotheses to social beliefs and practices. Modern western science, on this view, is a reflection of certain social ideas about rationality and instrumentality in relation to nature; science is useful for industrial societies that need to control nature and that is its primary justification and the principal reason for government and industry to support it. And, once this is admitted, if it ever is, the claim of science to a special status in relation to knowledge is likely to be challenged.

Thus, if there is a revolution in science today, it is a revolution in our conception of the nature of the scientific enterprise, leading to a clearer understanding that, far from being an exogenous activity, science has been (and is) moulded by the political, economic and social forces that characterize its environment. Seen in this context, the perennial problem of the social responsibility of the scientist takes on a new, more bewildering meaning; it is not simply a matter of scientists being concerned about the 'use and abuse' of their work, but of their actions in doing their work being part of an immensely complex and largely unpredictable system of social forces. But the idea of social forces already suggests that a certain patterning of events has taken place; that events do not simply follow one another without rhyme or reason, but may exhibit intelligible patterns. STS, then, has a triple role to play in contemporary society. The first is to examine critically the assumptions about the relationship between science and society which are guiding contemporary thinking and public policy. The second is to identify intelligible patterns of relationships — in other words, to explain how science and society are interconnected and evolve. Thirdly, its study provides a means for supplying our society with people educated in the complexity of the present situation of the scientific enterprise, and through this making it more likely that science will be directed more purposefully. Civil servants, journalists, industrial managers, trade union officials and politicians of every hue make daily use of scientific knowledge, but with what understanding, to what ends and for whom? Would it not be appropriate for more of the people who enter such careers to do so with an education in science or technology coupled with STS? Many leading scientists have thought so, but their views have so far had little effect on the teaching of science in secondary or tertiary education.

The structure of the book

It would be unrealistic to expect one book to do much more than set out some of the issues which are currently the concern of the STS community. Broadly, the chapters in this book are grouped around four different themes, though more than one theme is present in virtually every chapter.

(i) *The nature and origins of scientific knowledge*　The first section contains two quite different approaches to the problem of scientific knowledge. In 'The questioning of the scientist' (Chapter 1), an account is given of the pathways to the solutions of certain rather famous scientific problems. In examining these pathways the material is presented in a way designed to involve the reader in the process of historical reconstruction. This is an involved task and, hence, the chapter will repay careful reading and re-reading. Its purpose is to drive home the point that to understand what is 'going on' in the development of science both student and teacher need some first-hand experience of being caught up in a problem. Otherwise, the history of science is bound to appear somewhat abstract. Without attempting to grasp and grapple with the set of questions which occupied scientists during a given period, the student of STS is particularly vulnerable to the various *metatheories* of scientific development which are used to organize our understanding of the history of scientific development. Two such metatheories which are currently popular can be drawn from T.S. Kuhn's book *The Structure of Scientific Revolutions* and from Karl Popper's *Conjectures and Refutations* but, as the author of our first chapter argues, readers should pause and reflect before they assume that they know what these authors are referring to when they talk about scientific discoveries. It is suggested here that readers should try, in selected cases, to grasp the network of questions which have occupied other scientists and to which the historical record of theories are the answers. It is possible that even though the anwers may change, the originating questions might remain more or less constant within a given research tradition over long periods of history, and that significant new answers emerge when new questions, about, say, the nature of matter, or the origins of life, come to be formulated.

Still, once theories (answers) are formulated there is no guarantee that they will be accepted by everyone. It was Max Planck who observed that a scientific theory advances less through a process of intellectual conversion than by the demise of its opponents. But, between the enunciation of a theory and its eventual acceptance the possibility of controversy exists and it is to the nature of this that Chapter 2, on 'Authority versus argument in geology', is directed. It is often argued that within the scientific enterprise itself there is a hierarchy of disciplines, with mathematics and physics at the top and the other 'less mathematical' subjects ranked lower down. In one example given in this chapter, the interpretation of geological data concerning the age of the earth is presented as having had to cope with criticism from physicists who, using their then current knowledge of physical processes, reached a very different conclusion about the age of the earth. This, and the other cases discussed in Chapter 2, gives an insight into how geologists have had to strive to establish their right to scientific autonomy. Throughout the period being discussed geology appears to come off rather badly in the sense that it is dominated by the community of physicists, often because it could not provide a *mechanism* to explain the geological data. Clearly, there is more involved than this because in the final example, concerning the origins of the earth–moon system, there is a similar lack of an explanatory mechanism but apparently no controversy. Is part of the reason for this that by now geology has achieved its autonomy — that is, it is recognized as a science in its own right with its own data, theories, methods and techniques of measurement? If so, such controversy as arises would be more likely to take place *within* the community of geologists rather than come from another specialism.

(ii) *Scientific knowledge and political authority*   While the first section examines the origins and verification of scientific knowledge, the second explores some of the uses to which this knowledge can be put. *Use*, in this context, however, refers not to the application of scientific knowledge to industrial production but rather to the role that it may play in support or in subversion of political authority — for it can do either according to circumstances. On a conception of science which saw it as ethically neutral and as having emerged independently of social influences, one could imagine that science should be capable of resolving political disputes the substance of which was open to empirical investigation. We find, however, that very few burning political or social issues can be formulated for arbitration in this way. Moreover, if one does not accept the premises on which this conception of science is based, then one is even readier to accept that scientific data and arguments enter the political arena as one element in a highly complex process.

Chapters 3, 4 and 5 present widely different examples of the ways in which scientifically-validated knowledge has entered the political arena in relation to the scientific under-standing of intelligence, the conflict between the theory of evolution and the creationist account of the origins of life, and the role of expertise in the regulation of carcinogenic chemicals. In each case, scientific knowledge is being made to perform a dual role. Firstly, it is used to show that the position being argued is based upon valid methods, that is on methods that the relevant scientific sub-community would endorse, while refusing similar endorsement for other viewpoints. Secondly the knowledge is appropriated by another social group and used by them as part of a campaign to discredit opponents in the larger social or political process, while, at the same time promoting their own. Needless to say, when science is used in this way much of the nuance with which scientists qualify their arguments is lost and, as a result, scientific data often appear to be being used as an intellectual blunderbuss with which to destroy an opponent rather than as an element in a rational debate. When, in addition, scientists from different specialisms are seen to be competing for the mantle of scientific authority in relation to a given political question, the analysis of the debate will tax the most subtle minds.

(iii) *The economic impact of science and technology*   Chapters 6, 7 and 8 examine the relations between scientific knowledge and technology, the concept of technical change, and the impact of new technologies. Their aim is to deepen and qualify our appreciation that an industrial society is grounded in the industrial application of scientific knowledge. By looking historically at the interaction between science and technology it becomes clear that scientific knowledge is not *applied* in any simple way, but is involved with technical knowledge in a much more complicated process, the nature of which is changing. The interest in this problem is not merely historical, however. We are living in a period of major structural change characterized in part by the emergence of industries based upon new technologies. The development of these new technologies clearly depends upon many factors, but to the extent that scientific research influences that development it may be necessary for government and industry to bring the two activities into a closer relation, and so we must try to understand the nature of their interaction, and the interaction of both with the economy.

The chapters in this section show how important it is to be

critical of the methods used in the study of the interaction of science, technology and industry. To the extent that studies of the economic impact of science and technology themselves claim a measure of scientific objectivity, they are used in policy formulation to promote or hinder certain views about what society should be like. It is a central aspect of research in STS to try to bring to light the kinds of social preconceptions which underlie our understanding of the relationship between industry and society. If nothing else, such critical awareness provides some guarantee against naive promotion of technology as a cure-all for current or future social ills (or its damnation as their cause). The chapters in this section aim to show that, in confronting the complex problems of modern industrial economies, we must examine our assumptions about the relations between science and technology, and we must appreciate that technology is far from being the only formative influence upon industrial development. The future of industrial society is, in some complex way, ours for the making: it is not determined by the 'march of technology'.

(iv) *The control of science and technology*   The final section addresses more specifically the question of how societies may influence the direction to be taken by science and technology, though this theme is touched on more than once earlier in the book (e.g., Chapters 5 and 8). Through discussions of the politics of energy supply, the choices currently and soon to be made about the social acceptability of genetic engineering, and the problem of controlling the nuclear arms race, Chapters 9, 10 and 11 are concerned with the ways in which societies seek to manage the development and adoption of scientific and technological choices. Because of the importance of specialist knowledge in their prosecution, such policy areas reveal considerable tension between the roles of expert and non-expert, or generalist, in the decision-making process. The former group are right in feeling that decisions should be taken on the best technical advice available even though the majority of the population does not possess that competence. On the other hand, the population, in so far as it makes its worries manifest through special interest groups, is rightly concerned that decisions as to social welfare, the social acceptability of radiation hazards, and the ethics of nuclear deterrence, for example, should not be reached on the basis of technical considerations alone. So far, at least, the tension is unresolved but close study of the issues discussed in at least the first two of these chapters makes it clear that the problem is not only to resist the power of the technocrats but also to devise structures whereby broader social concerns can be effectively transmitted into the decision-making process. The final chapter pursues the somewhat different theme of the control of nuclear weapons, and demonstrates yet again the interpenetration of technical and political affairs, with the added complexity that the issue under discussion requires resolution not simply in the relatively calm waters of national politics but also in the turbulent and still largely uncharted deeps of international relations.

Where next?

The contributors to this book, in acknowledging the complexity of their subjects, have each observed that, in the space available, they can do little more than introduce the subject. This should not, however, be interpreted to mean that the problems presented could be solved if only more space were available: the limitations of space have restricted the possibilities for setting out the problems, rather than for answering them. Moreover, we have encouraged each contributor to be selective in his examples, and to present a particular way of viewing his problem rather than, necessarily, a fully rounded picture of it (though any necessary correctives may be found in the bibliographies and questions for discussion which follow each chapter).

There is, therefore, a dual sense in which this book is introductory. Firstly, it tries to pose a problem and indicate some of the methods which could be used in tackling it. Secondly, it draws on relatively simple examples or ones which require a minimum of previous experience on the part of the reader. Naturally, the presentation of any of this material in the classroom will require appropriate mediation by the teacher. Our aim has been not so much to produce 'classroom ready' teaching material as to provide school and college teachers, and other interested readers, with basic information, with some sense of how to approach each subject, with sufficient detail to move beyond the level of expressing mere opinions on the subject, and with some guidance via further reading and questions on how to press the inquiry further. While we cannot answer definitively many of the questions raised in these pages, we believe that the task of posing them should not be underrated. And if some questions are not raised at all (we are conscious, for example, that there are no chapters on food, population and development problems, on women and science, and on science and religion), this must be attributed to limitations both of space and of the expertise available in our group.

Happily, with the growth of Science Studies over the last decade, there now exists a number of centres teaching in this field, in the higher education sectors of several countries, to which interested readers could refer. These centres are characterized by a lack of consensus over their names (Science Studies, Science Policy, Technology Policy, Liberal Studies in Science, Science in Society, and Science, Technology and Society are some examples), but then one of the findings of STS is that such dissensus is typical of the emergence of new disciplines. Umbrella organizations such as the SISCON (Science in its Social Context) Project and the Science, Technology and Society Association have also arisen to develop teaching material, to act as focal points and to disseminate ideas. And in Britain and the Netherlands, at least, teaching material and examinable syllabuses (for example, the Science in Society Project of the UK Association for Science Education) have begun to be introduced into the schools. Thus there is now no shortage of opportunities for those interested to study further in this field and to expand not only their own intellectual horizons but also their capacity for engaging in the interactions between science and society.

Appendix

Further information on the field of Science Studies can be obtained from:

The Science, Technology and Society Association, c/o STSA Support Centre, Lipman Building, Newcastle-upon-Tyne Polytechnic, NE1 8ST, United Kingdom.

European Association for the Study of Science and Technology (EASST), c/o Professor J. Ziman, FRS, Imperial College, London, United Kingdom.

The SISCON Project, c/o Dr W.F. Williams, Director of Combined Studies, University of Leeds, United Kingdom.

SISCON-in-Schools, c/o Joan Solomon, Department of Educational Studies, University of Oxford, United Kingdom.

Science in Society Project, Association for Science Education, College Lane, Hatfield, Herts, United Kingdom.

In addition, undergraduate and postgraduate courses with major components in Science Studies may be found at many centres. Some examples are:

UK

    Ashton University (Technology Policy Unit)
    Edinburgh University (Science Studies Unit)
    Imperial College (Industrial Sociology Unit)
    Manchester University (Department of Science and Technology Policy)
    Sussex University (Science Policy Research Unit)
    Polytechnic of North London (Science and the Modern World)
    Chester College of Higher Education (Science in Society)

USA

Cornell University (Program on Science, Technology and Society)

Massachusetts Institute of Technology (College of Science and Society)

Canada

Montreal University (Histoire et sociopolitique des sciences)

Netherlands

University of Amsterdam (Science Dynamics)

Free University of Amsterdam (History and Social Aspects of Science)

University of Leiden (Chemistry and Society)

Further Reading

John Ziman, FRS, *Teaching and Learning about Science and Society* (Cambridge University Press, 1980).

A useful book which argues the case for educational development in this field.

F.R. Jevons, *Science Observed* (George Allen and Unwin, 1973).

Still a good general introductory text if you can find a copy.

H. M. C. Eijkelhof, E. Boeker, J. H. Raat and N. J. Wijnbeek, *Physics in Society* (1981: available in UK from SISCON, address above; in Netherlands from VU Bookshop, De Boelelaan 1105, 1081 HV Amsterdam).

An example of secondary level material produced for an examinable course on Physics in Society in the Netherlands.

1

THE QUESTIONING OF THE SCIENTIST: AN ENTRY TO SCIENCE STUDIES

Geoffrey Price

What are we doing when we are doing science? The purpose of this chapter is to suggest that if we — teachers or students of the sciences alike — are to make headway with that apparently divergent problem, we need an entry point: and that such an entry point is in our own experience, the curiosity that leads us to seek an understanding of the world. Consider first the biographical aspects of scientific curiosity.

For five years, Kepler struggled to account for Tycho Brahe's data on the orbit of Mars, feeling at times that a demon was knocking his head against the ceiling. His reader is led to the triumphant solution — only to be shown a fatal discrepancy of eight minutes of arc between observation and theory. Rather than abandon the attempt, Kepler persisted through six more years, nearly failing again from a series of errors, before stumbling on the secret of the orbit. Equally obsessive concentration is reported of other scientists. Newton's particular gift, said Lord Keynes, was that of holding a problem continuously in his mind for hours, days or weeks, until it finally surrendered its secrets. Einstein's dissatisfaction with the foundations of the physics he had been taught persisted through ten years of studying the original texts until he was able to clarify his own thinking sufficiently to write his first paper on relativity. J. D. Watson's personal account of the race to understand nucleic acid structures shows him, against his inclinations, learning new scientific disciplines and techniques to try to make headway with the problem, and finding that he could not dismiss the data from his mind even in relaxation.

Is this close involvement of the scientist in the puzzles posed by data something exceptional, or is it an element in every case of the growth of scientific understanding? What is the experience of *beginning* research like? Here are some comments of students at Massachusetts Institute of Technology, who had been able to combine working on 'live' research projects with their normal course work through a special Research Opportunities Programme.

I am getting accustomed to getting something wrong, and not being able to look up the answer in the 'back of the book' — very frustrating. At the same time, not having answers available anywhere is exciting — we are *writing* the back of the book, so to speak.

The most important thing I learned was the ups and downs and steady doldrums of research, what it feels like to get unexpected results which might prove fundamental in later research, or just to get tight-fitting data which support your theory ... The lab is so much more alive than the textbook.

(MIT Student Research Opportunities Directory, 1979.)

People with that kind of experience can tell us something of being 'drawn out of themselves' in the struggle to make sense of unexplored fields. Can we, however, learn very much about the growth of scientific knowledge by collecting and comparing these biographical and autobiographical accounts? Can we even understand what is reported, without some comparable personal experience to refer to? How can we evaluate the well-known examples we opened with: are they special cases, remote from the everyday activity of science, or do they simply exemplify scientific discovery in unusually dramatic form?

Experience of problem-solving

What fund of comparable experiences could we draw upon for comparison with biographical accounts of scientific discovery? No doubt some of the problems we have tackled in science or other fields have proved relatively easy, and they are not likely to have left detailed memories. By contrast, some will have been so daunting as to discourage us entirely, leaving little but the memory of bafflement. But some problems will have tantalized us with the fear that they may be just beyond our reach, and yet draw us on with the hope of finding a vital clue. What happened *then*?

That is the entry-point proposed in this chapter. Recall the times when puzzles that have interested you have begun to affect your pattern of living: eating, sleeping, daydreams, irritability. Recall the different tricks you tried for setting out the problem: the false trails, the sifting of advice for useful and false clues. Even if we lack the opportunity to become involved in live research programmes, the newer science syllabi of the last two decades provide opportunities to build up a fund of experience from which we may draw memories and reflections on the experience of tackling basic scientific problems. At the very least, our situation is probably better than when Einstein complained of his own schooling:

> It is nothing short of a miracle that the modern methods of instruction have not yet entirely strangled the holy curiosity of enquiry ... It is a very grave mistake to think that the enjoyment of seeking something and searching can be promoted by coercion and a sense of duty.
> (A. Einstein, 'Autobiographical Notes', in P. A. Schilpp (ed.) *Albert Einstein: Philosopher Scientist*, Evanston 1949, p. 17.)

Once we can identify within our memory the experience of becoming wrapped up in a problem, then we can go on to take the further step of asking, 'Are there any identifiable elements within the experience of problem-solving?' At this point, it will be helpful if we turn to an artificial problem as the focus of discussion, to reduce for the moment the distracting number of practical points in most real-life research. One such is the following:

> You are given twelve billiard balls of identical size, shape and colour. Eleven of the balls are of identical weight. The twelfth is *either* heavier *or* lighter than the rest. Using *only* *three* weighing operations on a set of scales, find which is the odd ball out, and whether it is heavier or lighter.

Now, if we are going to take seriously the suggested strategy of analysing our *own* experience of problem solving as an entry-point to understanding the wider questions of the growth of science, then there is no alternative but to try to tackle this problem before proceeding with the chapter.

What *is* its effect on your field of attention? The answer may be — none at all. It may seem too trivial, too artificial to engage you. In that case, it is clearly better to change to one that does, or to recall the confrontation with earlier problems that tantalized you. At all events, if we are to understand what is involved in the specialized type of curiosity that we call scientific questioning, it is not enough to read about the experiences of others reported in biographies. Indeed, as Karl Popper maintains, it may only be by trying to solve a live problem and failing to do so that we end up properly understanding what a problem is. If, however, your attention *is* diverted by this problem, you attempt trial solutions. Does it help to weigh six balls against six? Five against five? What are the implications of balance/imbalance? Note down as much as you can of the way that you encounter the problem: not only the content of the strategies that you undertake, but your encounters with bafflement, premature solutions, frustration and eventual relief.

What do you find concerning the type of attention that is required? Is progress related to formulating the problem clearly? The philosopher Robin Collingwood drew on his boyhood interest in archaeology to conclude that scientific investigation depended on being able to pose appropriate sub-questions in pursuit of the general problem which attracted him.

I found myself experimenting in a laboratory of knowledge, at first asking myself a vague question, such as 'Was there a Flavian occupation on this site?', then dividing that question into various heads and putting the first in some such form as this: 'Are these Flavian sherds and coins mere strays, or were they deposited in the period to which they belong?'

What, in your case, was the way in which you began to understand the consequences of the different weighing strategies?
What was involved in precisely assessing the effect of the
limitation to three, not four, weighings? Are there any parallels
to Collingwood's experiences? In retrospect, was there a period
of 'apprenticeship' before you clearly grasped the difficulty?
Would you then agree with Toulmin and Goodfield that

To analyse the difficulties facing you, stating them so clearly
and unambiguously that systematic observations or experiments can begin, is more than half the battle. Once that is
done, intellectual victory may take time, but it can be predicted with confidence; until then one's problems remain not
just difficulties but mysteries.

Perhaps you partly agree: but you may object that in your
experience the pursuit of blind alleys must also be recognized.
Not only did you 'automatically' approach the problem with
certain assumptions, but even when you reluctantly conclude
that you are stuck, you still are unable to conceive of any
other starting point. You may then be sympathetic with
Kepler, who spent a year struggling with a fruitless hypothesis
of an egg-shaped orbit for Mars, and even on eventually
breaking with it, still did not at first grasp that his new solution
had the simplicity of an ellipse. You may also have found
that different assumptions affect the way you set out a problem, and thus help or hinder your capacity to grasp the solution. With the billiard ball problem, algebraic summary
statements, or visual images of differently coloured or striped
balls may have been adopted, each of them embodying assumptions about the kind of solution you expect. To consider
giving up any of these approaches may well induce despair or
disorientation. J. D. Watson, already considering double-
spiral models for the backbone of DNA, could hardly bring
himself to consider the possibility that the intertwining of
the helices was through organic bases linked on the inside
rather than the outside.

Finally I admitted that my reluctance to place the bases
inside partially arose from the suspicion that it would be

possible to build an almost infinite number of models of this type. Then we would have the impossible task of deciding which was right.

(Ann Sayre, *Rosalind Franklin and DNA*, Norton Library, New York, 1975, p. 160.)

What of the final stages? Would you agree with Popper's remark that after we have failed a hundred times to solve a problem, we can be considered to have mastered it as a problem, and are able to tell immediately whether a proposed solution is correct? Even then, is it in any way fruitful to try to force yourself to find a solution? By now, the problem may have so focused your attention as to consciously affect your sociability, eating or sleeping. Yet the most significant steps may still seem to defy the will alone. J.D. Watson only discovered that he was using simplified textbook models for the backbone-linking bases in a chance conversation with a colleague. The testimony of Kekulé, obsessed with linear models for benzene, dreaming of snakes and thus of a closed-ring solution, is well known. Thus we find that problem-solving proceeds in an irregular and unpredictable fashion, beyond the reach of any systematic rules for discovery.

Understanding or seeing?

Our clearest memories are likely to be of the moment when the data of the problem, seemingly so divergent in its implications and resistant to understanding, begin to make sense. Historical examples in science are numerous: Poincaré and Darwin both recalled the exact time and place of major insights; Kepler felt that he had been awakened from a sleep when he stumbled by chance on the significance of a particular angle in the orbit of Mars. Hamilton, having been asked over breakfast by his two sons for a month whether he had succeeded in finding a rule for vector multiplication, was so elated at the solution he found while out walking that he cut the equation into the stone of the bridge he was passing.

Now what is involved in such moments of insight? Are we

simply 'seeing' the data in a new and unexpected way? Sudden reversals in our visual grasp — recall the well known diagrams that 'flip' from birds' heads to antelopes, from wine-glasses to facing silhouettes — have been much discussed as analogies to scientific insight, particularly in the wake of Thomas Kuhn's widely read *The Structure of Scientific Revolutions*. But *is* understanding quite like seeing — or touching, or hearing? You can only answer that question if you attend carefully to the different elements of your own awareness of tackling and resolving problems. Certainly, in common language, you may say with exasperation, 'Oh, now I see!' But *are* you seeing, or understanding? You may have a consciousness of understanding without that being a sensory consciousness. Try to recall as much as you can of the experience of understanding, from the memory of problems you have resolved.

What, though, of 'verification'? The debates in which Popper has accused Kuhn of treating scientific theories as inaccessible to empirical falsification have attracted much attention. In his *Conjectures and Refutations* and subsequently, Popper by contrast has tended to argue that to merit the status of science, hypotheses which are our hardwon insights must be subject to the most rigorous search for experimental refutation. The vigour of the debate (in I. Lakatos and A. Musgrave, *Criticism and the Growth of Knowledge*, Cambridge, 1970) makes it easy to neglect the ways in which Popper's presentation, just as much as Kuhn's, may encourage a conception of scientific insights as free-ranging mental constructs — albeit, for Popper, subject to the final veto of an 'out-there world'. Again, we may ask whether such models do full justice to the experiences involved in problem-solving. No doubt there are chains of sub-questions whose checking may be both important and simple. What though of the major inferences, in which we are forced to look back on the reliability and cross-connections of several lines of argument? For J. D. Watson, the regularities in the base ratios of DNA observed by Chargaff seemed for a long time vaguely relevant but unrelatable. Yet as his eventual solution moved from working argument to firm conclusions such data moved from the margin to the centre of his reflective awareness and began to make sense. In short, the mind of the scientist may not be so much inventing and *proposing* hypotheses, as *responding* to the intelligible relations that are to be found among the entities and events revealed through sense-data.

We can then identify distinct elements in the experience of tackling artificial problems – the experience of major and sub-questions, blind alleys, the grip of unrecognized assumptions, the progressive focusing of attention and energy, the absence of 'rules for discovery', the 'falling into place' of the evidence that allows the mind to pivot from the original data to express a new understanding, the slow reflection that reviews the sufficiency of the answers to problems within the evidence. How far, though, can we generalize from the experience of artificial problems to the structure of discovery within specific scientific problems? Truth, Collingwood insisted, is a property not of answers to minor questions alone, nor of a complex of such answers, but rather of a complex of questions and answers. We may have some sympathy with that view, after seeking to identify within our own experience the key role of clearly identifying the best strategic questions to pursue. Does it, though, enable us to move across to understanding the structure of scientific discovery, both individual and collaborative?

The biography of problem-solving: 'The Double Helix'

Watson's *The Double Helix* may be read as a document of personality clashes and motivation in science, or simply as a narrative of a major discovery, and it then invites comparison with Anne Sayre's contrasting account, *Rosalind Franklin and DNA*. Suppose, however, we consider it not simply as narrative, but as a progressive clarification of the major question that challenged Watson, and the complex of sub-questions and answers through which it was clarified? If we do, then it is quickly apparent that such a biography cannot be taken in isolation. Indeed, since the work of Mendel in 1865 on variability in pea plants, the question of heredity had been slowly coming into focus. Watson, interested since student days in the developing notion of genes as the unit of inheritance, first began to move into the field when many years of work had left researchers divided as to whether genes were special types of protein molecules, or deoxyribonucleic acid (DNA) found in the chromosomes of all cells. Moreover, the concerns of geneticists and of those organic chemists working on DNA were divergent to a degree that Watson found intensely discouraging (Watson 1970, 28–34). Even when taking time out from other work to attend a discussion meeting, Watson found it impossible clearly to formulate a question about gene structure until his attention was attracted by the work of the King's, London, group on the crystallography of DNA. 'Now, however, I knew that genes crystallize; hence they must have a regular structure that could be solved in a straightforward fashion.' (p. 35)

Yet if Watson was taking up the major question at a point where much work had already been done to clarify it, was he yet aware of the intellectual skills and interests he would need to muster in order to make further progress himself? Attracted by Pauling's work on helical structures in proteins, he began to make plans to investigate whether DNA was also helical (pp. 37–42). In a sense, he was beginning to clarify a sub-question – but without even an elementary knowledge of X-ray diffraction by crystals (p. 41). These signs of 'apprenticeship' continued for many weeks. Even when listening to new work from London on the crystallography of DNA, Watson failed to notice crucial details in the report (p. 65). In trying to construct a helical model, his ignorance of text-book stereochemistry became apparent (p. 69). In the arguments with the London group that were involved in questioning the adequacy of their first insights, it became apparent that Watson's model did not meet all the relevant questions that could be raised (pp. 78–81), and thereby resolve the basic issue through the coherence of a network of questions and theories.

The signs of mastery of the problem as problem were equally long in emerging. Casually, while still discouraged, Watson started consulting Pauling's *The Nature of the Chemical Bond*; he mastered X-ray diffraction sufficiently to prove the existence of a helical structure in viral material (pp. 100–1); Chargaff's data began to attract new attention as the problem began to focus (p. 102). Even so, the problem remained intractable. It would be instructive to make your own reconstruction of Watson's attempt to reconcile helical models, the X-ray data, and the question of the position of the organic bases, as described in his book on pages 130–48. What is the role of the different strands of data — chemical, crystallographic, structural? Where does the key insight occur? Where does it give way to judging the truth of the new possibility? Were Watson and Crick deliberately seeking ways of falsifying their model, or did they display an informed sense of the points at which all the chains of inference and evidence would have to concur for their theory to be reliable?

Problem-solving in the growth of discipline

Up to now, our personal and biographical evidence has suggested that there are (often lengthy) processes of apprenticeship and maturing to be surmounted before there is any prospect of grasping the full meaning and ramifications of scientific evidence. What then of the maturing of problems to which no complete solution is accessible in the lifetime of those who first opened them up? Is it possible to detect in the establishment of whole scientific *disciplines* just such transitions from apprenticeship to clarification of the central problems, on to the opening of fruitful lines of questioning and the beginnings of the resolution of problems? To ask that question seriously will expand our investigation well beyond the scope of this chapter. Yet it is not simply a question of historical interest. Extending Collingwood's model, we may ask, to what extent are the language and theories of today's scientific disciplines themselves the expression of answers to past questions? Do we fully understand today's theories if we do not grasp the points at which they succeed in resolving those central questions? Recall the MIT student who had to get used to 'not looking up the answer in the back of the book'. That is the discovery of the difference between the capacity to manipulate existing answers to previous questions, and the skill to pose pertinent and well-timed questions that allow new answers to be formulated. Equally, we can ourselves experience the contrast between accepting scientific terms through authority or the learning of routines, and understanding the significance of those terms as responses to questions.

Consider the familiar language of the molecular hypothesis. Recent work on John Dalton has made it possible to appreciate his slow realization that his attempt to develop an atomic theory of the physics of gas mixtures might also provide a clue to the problem of chemical combination. From that point, Dalton pursued his clue with ferocity (even though within a few years his later theory of gases had undermined the original argument!). Thus he publicly argued that

in the Chemical Union of Elementary Principles, we shall generally, if not always, find a compound, consisting of one atom of each. That the next most simple combination is 1 atom with 2. That such combination always take place before more complex ones, which in reality do rarely occur
(Dalton, *Prospectus* for 1807 Lectures.)

Thus Dalton took the large-scale observation of simple numerical ratios of combining weights when gases reacted to form alternative products (e.g., nitrogen + oxygen → nitrous oxide, nitric oxide, nitrogen dioxide) to be a clue to simple geometric combinations of their atoms in 'reaction neighbourhoods'. Further, he defended his view that the simplest (A–B type) compounds would always be formed first because he believed that two similar atoms could not be present in a reaction neighbourhood. Only subsequently could ABA or BAB compounds be formed by A + AB or AB + B combinations, and even then like atoms would not be adjacent.

Criticism came from several angles. The strict experimentalists, concerned with the practicalities of combining weights, saw the ambiguities in Dalton's theory: if A and B only formed two compounds, how could one tell which was the simpler? Defenders of non-corpuscular theories of matter found Dalton's tenet of many, element-specific types of atoms unnecessarily complex. Humphry Davy, prominent among them, criticized Dalton's far-reaching inferences even in a speech presenting to Dalton the Royal Society's Royal Medal. But is this not now long forgotten? In any case, might not a corpuscular theory have found comfort in Gay-Lussac's results of 1809 showing simple numerical ratios in the combining *volumes* of gases as well? Might not 2 : 1, 1 : 1 and 1 : 2 volume ratios in the formation of the nitrogen oxides imply parallel relations in the ratios of combining atoms? But does that answer all the relevant questions? What of other gases?

Gay-Lussac's volume data for steam, ammonia and nitrous oxide were:

2 volumes hydrogen + 1 volume oxygen yields 2 volumes steam
3 volumes hydrogen + 1 volume nitrogen yields 2 volumes ammonia
2 volumes nitrogen + 1 volume oxygen yields 2 volumes nitrous oxide.

Dalton's arguments force him to represent steam as HO, ammonia as NH and nitrous oxide as NON. *He* would therefore represent the three reactions as

(1) H + O = HO
(2) H + N = NH
(3) 2N + O = NON

Suppose that equal volumes of gases contain equal numbers of Daltonian atoms. Then *Gay-Lussac*'s data would translate as:

(4) $2H + O \rightleftharpoons 2HO$
(5) $3H + N \rightleftharpoons 2NH$
(6) $2N + O \rightleftharpoons 2NON$

Clearly the corpuscular theorist is in difficulties. But why should equal volumes be assumed to contain equal numbers of atoms? Take the data on steam: if both hydrogen and steam only contain $0.5n$ atoms per unit volume, whereas oxygen contains n, then the volume data translates as

(7) $H + O = HO$

giving a balance. But the price of resolving conflict in this way is that we maintain the same assumptions throughout. What of the nitrous oxide data? Can you find a conversion factor for nitrous oxide that will balance the translation to atomic terms? If so, can the same be done for ammonia formation? Does this line of enquiry lead into a blind alley? Perhaps it is hopeless to assign different volumes conversion factors for every elemental and compound atom. Should the original assumption that equal volumes contain equal numbers of atoms be reconsidered? But is that possibility not blocked by the lack of balance in equations (4) − (6)? Suppose, however, that we combine the equal volume/equal number postulate with another: that 'atoms' of gaseous elements are not necessarily single, indivisible Daltonian atoms, but may themselves contain more than one atom. Then we can assign trial 'molecular' formulae to hydrogen, nitrogen, oxygen, ammonia and steam in turn, and satisfy ourselves whether the problem can be solved. Our appreciation of 'live' problems will enable us to enter the questioning underlying this classic issue. Indeed we may ask, had the problem reached resolution with this, Avogadro's famous proposal? What view would a strict Daltonian have to take of 'molecules'? What would a 'molecular' theorist feel about the later discovery that bromine and iodine did not strictly obey the gas density laws at high and low temperatures? Did Avogadro provide an unambiguous basis for defining molecular formulae?

We might here embark on a project of enquiring whether the chemical community in 1810—20 was at the stage of apprenticeship, mastery or resolution in dealing with the structure and intercombination of matter. But indeed we might ask the same question concerning the central questions of our own discipline today. And then we can really start to appreciate what is involved in 'writing the back of the book'. Conversely, if we do not appreciate the central problems posed by the evidence of our field, whether in the physical, the biological or the psychological sciences, do we really understand what we are doing when we are doing science? If we have not shared in the attempt to develop an understanding of puzzling data, how can we even begin to consider whether scientific theories are ultimately our own inventions, or attempts to express the intelligibility of the universe?

Questions for discussion

1. To what extent are the elements of the discovery process you have identified unique to science?

2. Compare and contrast the process of discovery in Kepler's work on the orbit of Mars, and Watson's search for the chemical structure of DNA.

3. Compare the chain of question and inference in the final stages of the Double Helix investigation, with Collingwood's illustration of enquiry in the model detective story, 'Who killed John Doe?' (*The Idea of History*, Oxford 1939, pp. 269—282). As you examine Watson's account of the move from the last blind alley to a resolution of the problem, can you identify any similarities with Collingwood's account of reflective, 'verifying' questions? Drawing examples from your analysis of the questioning process in science, consider whether science is more concerned with 'inventing' the world, or with understanding it.

Further Reading

W. I. B. Beveridge, *The Art of Scientific Investigation* (Heinemann, 1961).

Includes a wealth of descriptive material on all aspects of the experiences of the discoverer.

James D. Watson, *The Double Helix* (Penguin, 1970).

A dramatic if idiosyncratic first-hand account of the crucial steps in the Cambridge investigation of DNA.

Arthur Koestler, *The Sleepwalkers* (Penguin, 1964).

A pioneering reconstruction of crucial period in the emergence of modern science, focusing on personal experiences of Copernicus, Kepler and Galileo.

Arnold Thackray, *Atoms and Powers* (Harvard University Press, 1970).

A scholarly reconstruction of pre-Daltonian chemistry and its critics. Only for those with detailed interest in this period as a case-study.

W. Mathews and G. L. Price *The Nature of Scientific Discovery* (Science in a Social Context Project, 1979).

Full text of thirteen-lecture course including historical case study and collection of primary papers. Designed to make possible the approach to the discovery of scientific curiosity that is outlined in the present paper. (Available from Dr W. F. Williams, Director of Combined Studies, University of Leeds.)

2

AUTHORITY VERSUS ARGUMENT IN GEOLOGY

Glyn Ford

Introduction

A fashionable question in social science examinations a few years ago took the form: 'Is History a science?'; 'Is Economics a science?'; 'Is Sociology a science?'. The question arose at a time when the social sciences sought the *cachet* that was granted to 'sciences'. More recently this type of question has lost popularity: a devaluation of 'science' has occurred and enthusiasm for admission into the club has waned.

This devaluation has been brought about by growing awareness of the work of historians and philosophers of science over the past two decades, especially Thomas Kuhn's book *The Structure of Scientific Revolutions*. But before considering Kuhn's work it is necessary to look at what is implied by the term 'science' as it was used above. Many alternative definitions of 'science' are available, but in the context of pre-Kuhnian thought, these all contained — implicitly or explicitly — a shared and interlocking set of preconditions that *at minimum* had to be adhered to in practice if a subject was to be classified as a 'science'. These preconditions required that statements about the world were objective, and derived from a rigorous, logical and open examination of all the available data (and *not* based on work that was partial, prejudiced or illogical). The work of the scientist was done without fear or favour. Scientific statements were based upon evidence which resulted from research that took place independently of extra-scientific (external) considerations.

In contrast, non-sciences made statements about the world that were subjective.

Kuhn cast into doubt this claimed sharp distinction between science and non-science. He studied in some detail a number of key periods in the history of science when fundamental new theories swept aside others that had served for decades, if not centuries, as frameworks for scientific activity. On the basis of this work, he claimed that at such revolutionary periods in the history of science the choices between competing theories were made in the last analysis on *unscientific* grounds, or, as he puts it, relying on 'more subjective and aesthetic considerations'. This does not mean these choices are entirely arbitrary, though some of those who have continued Kuhn's work have been prepared to argue more bluntly that 'science is socially determined'. But even without going this far, Kuhn's work threatens to destroy the traditional edifice of science. While few would contest the argument that art, literature, music and law reflect the social values of the milieu and the interests of the individuals that brought them forth, science's key claim to a superior status relies upon its objectivity, its independence from societal influence. If Kuhn is right, then science's position within our society has been obtained with false references.

It is not my intention to present the full Kuhnian case here. Rather, I intend merely to show, using examples from geology, that the earlier claims of objectivity made for science are not always met in practice. More particularly, I intend to show that the ideal of 'open examination' of the data is often replaced by reliance on authority.

For the sake of clarity, it is worth emphasizing that I do not suggest that the individual scientists who figure later in this chapter, such as William Thompson (Lord Kelvin, as he later became) or Sir Harold Jeffreys, were anything other than perfectly sincere in their beliefs and, proper in their scientific practice. True, they were both exceptional men of their period and their profession; nevertheless they were still part of both, and reflected values, scientific and otherwise, accordingly. Their work was not 'bad' science or pseudo-science: had this been the case, my argument would collapse, for it depends upon the premise that the examples used here are not untypical of good science.

Obviously, the case made here cannot be assumed to apply to all sciences, and the claim could well be made that geology is different from other sciences. What I hope to show is that at least one science fails to conform to the traditional 'rules' of science. Perhaps, too, this chapter will stimulate wider interest in examining the social context in which scientific knowledge is produced and thereby increase our understanding of the extent to which contemporary science may be said to be objective and value free.

The age of the Earth debate

Today the Earth is believed to be between 4,500 and 5,000 million years old. These figures are not particularly controversial (except among Creationists, see chapter 4). In the mid to late nineteenth century this was not the case. To understand why this was so requires some appreciation of developments in natural history at the time.

Prior to 1830, the history of the Earth was explained on the basis of a series of individual catastrophes: gigantic convulsions, floods and volcanic eruptions. Then, in this year, the geologist, Charles Lyell, published his *Principles of Geology*, in which he suggested that the whole of Earth history was explicable on the basis of one simple principle, 'uniformitarianism'. He held that the slow and steady operation of recognized present-day causes was alone responsible for the raising of mountains and their subsequent disappearance. The present was the key to the past.

It was a bold suggestion. The summit of Mount Everest is marine limestone. Lyell — although he could not know it, for Everest was unclimbed — was suggesting that these and similar rocks had made their ascent of about thirty thousand feet not by a catastrophic convulsion, but by continual, imperceptible upthrusting over long periods of time. Lyell's principle had a simplicity and elegance that quickly found favour among younger geologists. No longer did they have to provide singular explanations based on a selected cocktail of catastrophes for each particular conjunction of rocks; instead they could invoke the general principle of uniformitarianism as an explanatory tool.

Lyell's friend, Darwin, in publishing *On the Origin of Species* in 1859, proposing the evolution of species by natural selection, analogously suggested that slow, imperceptible change could, over long periods of time, effect similar transformations among the fauna and flora of the Earth to those which Lyell had proposed for the lifeless rocks. Now, in terms of the theory of catastrophism the age of the Earth was an almost irrelevant side issue: the required number of catastrophes could be fitted into almost any time period. This was not the case with uniformitarianism and evolution. They both required immense passages of time for their twin principles to operate. As Darwin wrote in *On the Origin of Species*, 'he who can read Sir Charles Lyell's grand work on the Principles of Geology and yet does not admit how incomprehensibly vast have been the past periods of time, may at once close this volume'. It was this challenge of Darwin's, and the response to it of one great physicist in particular, that caused geology much difficulty in the ensuing decades. That physicist was Lord Kelvin.

Kelvin's interest in the age of the Earth pre-dated Darwin's *Origin* but he had not seen that the uniformitarians' demands for time might conflict with his own beliefs based on his scientific work. This was partly because Lyell had not quantified his demands, expressing his requirements in terms of 'indefinite' or 'limitless' expanses of time. This could be read in a number of ways and was generally taken to mean something far in excess of Archbishop Ussher's well-known estimate that the Earth was 6,000 years old. Darwin was more explicit than Lyell, suggesting that the erosion of the Weald alone would have taken 300 million years. This Kelvin found unacceptable.

He was encouraged to make this clear because, although not opposed to evolution as such, he did reject natural selection because he believed that it left no place for the operation of design or divine order in the evolution of life. Kelvin's theological beliefs therefore spurred him to confront both uniformitarianism and natural selection with the results of his physics. From 1862 onwards he did this in a series of papers.

Kelvin rested his case on three lines of argument. First, he examined the source of the Sun's heat. Arguing that since a finite body cannot be an infinite source of energy, nor a constantly replenished chemical or mechanical source of supply, he searched for present-day actions which could account for what was known of the Sun's past radiation and present temperature and mass. After investigating a number of alternatives he claimed that one cause fitted all the requirements. He dismissed chemical action, as even on the basis of the most energetic chemical process known an object of the mass of the Sun could only generate heat for 3,000 years, and suggested instead that meteoric action could supply heat for up to 20 million years. The theoretical argument behind this was that if a large number of meteors of mass equal to that of the Sun were drawn together by mutual gravitational attraction, their original potential energy would be progressively converted into heat energy to form the incredibly hot molten mass of the Sun, which would slowly contract over time, releasing heat as it did so.

Kelvin's second argument was to apply the same reasoning to the Earth itself. He imagined that the Earth had originally been a molten mass and cooled over time to its present internal temperature. In this case he was able to use in his calculations actual measurements of certain parameters that in the Sun's case he had had to estimate. The result was an age for the Earth of 98 million years, but to allow for uncertainties he gave upper and lower limits of 20 and 400 million years.

Kelvin's third argument was based on the fact that tidal friction — details of which are given later — slows the Earth's rate of diurnal rotation. The Earth's shape will indicate, by the degree of flattening it shows, its angular velocity at the time of solidification. If the rate of retardation is known then its age is calculable. Here. because of uncertainties, Kelvin was prepared to go no further than say that the Earth's shape was consistent with solidification 100 million years ago.

Kelvin did not stick rigidly to the results of these calculations. He was prepared to modify them on the basis of criticism and new knowledge, and this he did over nearly three decades. This compounded the geologists' problems for, as a result, they faced a progressively restrictive chronology. In 1862 Kelvin was prepared to admit at least the possibility of a 400 million year old Earth, but by 1868 this had been reduced to 100 million years, by 1876 to 50 million years, and by 1897 to 24 million years.

The response of the geological community was not what one might have expected. They accepted the authority of Kelvin's physics to set unchallenged limits on geology. They were either convinced or cowed by Kelvin's mathematical treatment of known physical phenomena, and adjusted their speculations accordingly. Sir Archibald Geikie, one of Britain's foremost geologists, accepted in 1868 that 'about 100 millions of years is the time assigned within which all geological history must be comprised'. He was followed by Alfred Russel Wallace, the co-discoverer of natural selection, and by almost everyone else. There were geological and geophysical arguments against Kelvin, but these had to be made by geologists far outside the main stream and their arguments fell upon deaf ears.

One counter-argument was put by Samuel Haughton. In 1864, adopting a different rate of secular cooling from that assumed by Kelvin, and claiming that past changes in climate have followed the loss of the Earth's internal heat, he estimated the Earth to be well over 2,000 million years old. This failed to encourage the geologists to look at Kelvin's assumptions.

A second counter-argument was based on what was then the most common geological method of estimating the Earth's age, namely, the use of the rates of sedimentation or denudation as timekeepers. This involved calculating both the total aggregate thickness of sedimentary rocks and the average rate of deposition or erosion. A large number of simplifying assumptions needed to be made, as with Kelvin's own estimates. One estimate was made by T. Mellard Read in 1878, based upon the soluble mineral content of the rivers of England and Wales as a way of estimating denudation rates. The result was an approximate *minimum* age of 526 million years. But both of these geologists repeated their work in later years, arriving at results consistent with Kelvin's 100 million year estimate. By 1880 even this limited and marginal geological work on the age of the Earth had stopped.

By the 1890s, the continued tightening of the temporal belt began increasingly to discomfort the geologists. Even Geikie felt compelled to argue in 1895 that

there must somewhere be a flaw in a line of argument which tends to results so entirely at variance with the strong evidence for a higher antiquity, furnished not only by the geological record, but by the existing races of plants and animals. They have insisted that this evidence is not mere theory or imagination, but is drawn from a multitude of facts which become hopelessly unintelligible unless sufficient time is admitted for the evolution of geological history. They have not been able to disprove the arguments of the physicists, but they have contended that the physicists have simply ignored the geological arguments as of no account in the discussion.

In 1899, Geikie acknowledged the failure of geologists to put their case over the previous thirty-five years and asked them to put their arguments on the sound basis of measurement and experiment. However, it was not until the discovery of radioactivity, and the consequent availability of new explanations for the source of the Sun's energy and the internal heat of the Earth, that geology was able to free itself from the constraints of the physicists' original timescale.

It is difficult to fault Kelvin for his work, which lay firmly within the assumptions of nineteenth-century physics. The problem lay with the geologists, who baulked at following the logic of their own discipline, opting instead to subordinate themselves to physics in general and Kelvin in particular. The result was that geologists performed intellectual acrobatics in an attempt to compress Earth history and natural selection into the timescale that the physicists had allowed them.

Alfred Wegener and continental drift

Plate tectonics has been the fundamental organizing principle of geology for at least the past decade, yet in its essentials the theory is almost seventy years old. It was first proposed by Alfred Wegener, a German meteorologist and geophysicist, in 1915. His theory was published in *Die Enstehung der Kontinente und Ozeane* (*The Origin of Continents and Oceans*), where he detailed his reasons for believing in lateral movement of land masses. In particular, he emphasized the fact that Africa and South America were at one time joined.

The geological evidence in favour of his theory was impressive. Wegener argued that the pattern of rocks in South Africa matched those of Brazil and Argentina. As he wrote, 'it is just as if we were to refit the torn pieces of a newspaper by matching their edges and then check whether the lines of print run smoothly across. If they do, there is nothing left but to conclude that the pieces were in fact joined this way'.

Wegener's theory also fitted in with the palaeontological evidence. It had been known for a long time that striking similarities were to be found in parts of the fossil record of continents that were now separated by distances of thousands of kilometres. Similarities between Brazil and Africa, Australia and Africa—India, and South Africa— Madagascar and India, were all particularly strong. Previously, the explanation of these similarities had been that the halves of each pair were connected by transoceanic land bridges that had subsided without trace after the Cretaceous, about 70 million years ago. The beauty of Wegener's theory was that it swept away the necessity for these rather improbable temporary land bridges, and instead suggested that the pairs had been directly connected, with the continents joined, during the relevant geologic period.

Palaeoclimatology could be rescued from an even worse predicament. Today a number of climatic zones roughly paralleling lines of latitude can be distinguished. These zones are not found in the geologic record. For example, Europe's climate had changed in the period between the Carboniferous (about 300 million years ago) and the present from tropical to temperate, while Norway's (e.g. the Spitzburgen archipelago) had changed from sub-tropical to polar. These changes cannot be explained merely by suggesting that the overall climate of the Earth had been hotter. Evidence of the presence of ice-sheets during the Carboniferous can be found in India and Brazil, as little as 10° south of the equator. There is also evidence that what is today the northern United States had a tropical climate in the Carboniferous period. These departures from the present symmetrical climate zones north and south of the equator were an embarrassment to geologists; but with mobile continents the anomaly vanishes. Yet despite the fact that Wegener's theory was able to tie together and make explicable all these otherwise disconnected and unaccountable observations, there was a problem. This was the absence of any convincing mechanism that would allow drift to occur. Wegener did suggest mechanisms in his book. He quoted the geophysical evidence that the substratum beneath the Earth's crust can act as a fluid allowing *vertical* movement and suggested that lateral movement was also possible and took place because of the action of two separate forces. The first of these was tidal in origin and arose because the retardation of the Earth's diurnal rotation by tidal friction occurred unevenly. Because the continental fragments were partly decoupled from the underlying layers and because tidal friction concentrated its effects on the outermost layers of the Earth the result was that the land masses slid over the substratum beneath the crust. Wegener's second force was the *Pohlflucht* force, attributable to the fact that the Earth was in reality not a perfect sphere but an oblate spheroid. The differential gravitational force exerted because of this variation in distance from the centre of the Earth led to a slipping away from the two poles.

It was some years before Wegener's ideas became known in the English-speaking geological community. When they did, in the early 1920s, it was the geophysicist Sir Harold Jeffreys who played Kelvin to Wegener's Lyell. Jeffreys rightly demolished Wegener's two forces in early editions of his book *The Earth*. He showed, for example, that if the tidal forces were strong enough to cause westerly drift of the Continents on the scale required then the Earth's rotation would be arrested completely within twelve months, but at the same time he dismissed the geological and palaeontological evidence for drift in a few short sentences. Again the response of the geologists was in general to ignore the theory which threatened to be so helpful to them. A few expressed some interest but even this was killed when, at a conference held in 1926 by the American Association of Petroleum Geologists, the whole idea was roundly condemned by a succession of eminent geologists in Wegener's presence. Jeffrey's treatment of him was gentle compared to what he had to endure from some of his other critics. Many of these went as far as to call into doubt his respectability as a scientist. Two, not untypical, quotations will serve to illustrate these attacks:

Whatever Wegener's own attitude may have been originally, in his book he is not seeking truth; he is advocating a cause, and is blind to every fact and argument that tells against it. Much of his evidence is superficial. Nevertheless he is a skilful advocate and presents an interesting case. (P. Lake)

[Wegener's method] in my opinion, is not scientific, but takes the familiar course of an initial idea, a selective search through the literature for corroborative evidence, ignoring most of the facts that are opposed to the idea, and ending in a state of auto-intoxication in which the subjective idea comes to be considered as an objective fact. (E. W. Berry).

The idea of Continental drift sank into oblivion for the next thirty years. Geologists failed to argue their own case, and so stood mute as they could not argue on the ground chosen by physicists. Jeffreys, within the context of physics in the inter-war period, was correct to dismiss Wegener's mechanisms, but the geologists by their silence surrendered an idea that promised to put together many otherwise isolated pieces of the geological jigsaw. Its revival in the mid 1950s owed much to wartime oceanographic work that had been undertaken for reasons far removed from those of pure research.

The origins of the Earth–Moon system

Our previous examples have been historical. The issues they were concerned with have now been resolved, in so far as any issue in science is ever finally settled. Contemporary examples are more difficult to identify because of the very fact that the issues are still in doubt. It is not possible to identify any major controversies in the geological sciences on a par with those already examined, but in those marginal areas where the work of geologists and physicists overlaps, cases do seem to exist. A clear example is the issue of the origin of the Earth–Moon system. Here work has been done by both astronomers and geologists. Their results seem contradictory, but none of those working in the area seems concerned to resolve the contradiction.

The Earth is unique in the solar system in having a satellite of such size that it exerts considerable forces upon its parent body. The tides raised on the Earth are a part of this phenomenon. These arise in the oceans and within the solid earth. They also have an effect on the dynamics of the Earth–Moon system. (This effect was mentioned above in relation to Kelvin's calculations of the age of the Earth.) To explain how this is caused it will help to look at Figure 1. Figure 1 (*a*) shows the situation that would obtain if the Earth were perfectly elastic and the oceans perfectly fluid. This, however, is not the case and the situation illustrated in Figure 1 (*b*) actually applies. Because the Earth is not perfectly elastic and the oceans are viscous, the maximum height of the tides does not lie along the direction of the attracting force. The rotation

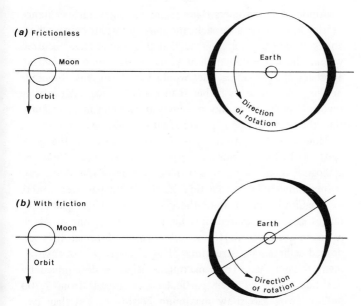

Figure 1 Schematic representation of the mechanism causing tidal friction

of the Earth carries the bulge forward; and the tide is high not when the Moon is directly overhead, but at some time later. (This may not seem obvious but is, nevertheless, the case.) The gravitational attraction of the two bulges is therefore off the line of centres between Earth and Moon. At the same time there is a stronger forward attraction from the bulge nearest the Moon than a retarding one from the bulge on the other side of the Earth. This gives rise to a torque on the Earth and an equal and opposite one on the Moon. The result is that the Earth's diurnal rotation is slowed and the moon is hurried on in its orbit, in consequence spiralling very slowly out into space.

Astronomers interested in the dynamic history of the Earth–Moon system have extrapolated backwards the effects of these forces on the system. Their calculations indicate that at some period in the past a close approach of the Moon to the Earth took place.

Amongst astronomers there seems to be general acceptance of this close lunar approach, the only questions being, when, and how close? All are agreed that this should have occurred within the last 3,000 million years. At the time of such an approach the prevailing tides would have been tens, hundreds or thousands of times higher than those of today. Alternative theories, such as the Moon joining the system more recently than the time of close approach, find little favour.

However, if we turn from the astronomers to the geologists a difficulty appears. There is no evidence within the geological record for a catastrophic capture of the Moon, nor is there any evidence that tidal heights in the past were orders of magnitude higher than those of today. These paleotidal ranges can be measured in a number of ways, one of which is to measure fossil stromatolite heights. Stromatolites are domed sedimentary structures of algal origin whose environment is inter-tidal. Their maximum height is determined by the tidal range of their particular site. Fossils from 2,700 million years ago show maximum heights of less than two feet, implying that tidal ranges were of the same order as today. The same is true of other measures of tidal range. Tidal range evidence is available from as far back as 3,200 million years ago, and no measure of tidal height between then and now fits that required by the astronomers.

The two communities seem to have produced contradictory results. In this case, unlike our previous examples, neither community seems particularly disturbed by the contradiction.

Conclusion

In the space available it has not been possible to do justice to the scientific complexities surrounding my three examples. (The reading list gives details of further sources for each example.) The examples are not identical in the lessons they offer. The key point is that they all show in their varying ways how both geologists and physicists have failed to maintain the open-minded, constantly questioning approach that is traditionally claimed to be the ethos of science.

The reasons for this are many and there is space to outline only some of them. The main cause seems to be an undue respect for *authority* in a number of guises. First, there was to some extent reliance on the *personal authority* of charismatic individuals like Kelvin and Jeffreys. More important, however, was *mathematical authority*. Science has within its extended family of disciplines a developed hierarchy of subjects. Those sciences that are more 'exact', more mathematically based, are regarded as superior to the 'softer' descriptive sciences, and all science aspires towards the standards of physics. While this remains the case, geology seems doomed to play second fiddle to physics. The calculations of the physicists have a strength missing from the qualitative evidence of geology. Certainly, Kelvin's own view was that data that cannot be quantified are barely worthy of a scientist's attention. This perception is reinforced when geologists themselves attempt, with at best partial success, to emulate the 'exact' sciences in their discipline.

Finally there is the *authority of ideas*, especially in the shape of the primacy of theory over observation. This can be particularly seen in the first example where the strong theoretical backing for Kelvin from physics overwhelmed the limited attempts that were made to challenge them with the semi-quantitive data available to the geologists. In our final example, the myopia of the physicists in failing to consider the geological evidence bearing on the history of the Earth—Moon system is aided and abetted by the absence of alternative theories which might explain the geological data.

There is no blame to apportion for what is outlined here. I have simply offered some observations on how a science develops which are in contrast with the prescriptive guidelines of the pre-Kuhnian philosophers. The awe which geologists showed towards the physicists is a product of placing authority before argument.

As a final note, let us observe that individuals are not

impartial. In Kelvin's case the enthusiasm with which he pursued the hapless geologists over the question of the age of the Earth owed something to his religious predilections. What is disturbing is that an individual's views and opinions may pre-empt the scientific debate, or that when debate does take place these may be given undue weight because of his position or his discipline. In Wegener's case — as we have seen — personal abuse sometimes substituted for scientific argument. Yet in none of these examples — apart possibly, from the first — were passions particularly charged. How much more creaky were the traditional norms of scientific practice when they had to bear the weight of passion aroused by the Race—IQ debate, or when dealing with the effects on children of low-level radiation?

Questions for discussion

1. Does the evidence here show that the 'traditional' definition of what constitutes a science is invalid for geology?

2. Why does geology subordinate itself to physics? Is this something peculiar to geology, or is it true of other sciences?

3. How should scientists modify the claims they make about their practices in the light of the argument of this chapter?

4. Lord Rutherford claimed that: 'All science is either physics or stamp collecting'. Is this valid?

Further reading

Introduction

T. S. Kuhn, *The Structure of Scientific Revolutions* (Chicago University Press, 1962).

For a contrasting view on the philosophy of science see some of the works of Karl Popper. A good introduction is available in B. Magee, *Popper* (Fontana, 1973).

The age of the Earth

J. Burchfield, *Lord Kelvin and the Age of the Earth* (Macmillan, 1975).

Alfred Wegener and continental drift

A. Hallam, *A Revolution in the Earth Sciences* (Clarendon Press, 1973).

A Wegener, *The Origin of the Continents and Oceans* (Dover, 1977).

Scientific Revolutions (Open University Press, 1981). This is units 12–13 of the Course U202 *Inquiry* for the Open University.

W. Glen, *The Road to Jaramillo* (Stanford University Press, 1982).

The origin of the Earth–Moon system

Z. Kopal, 'The Earth–Moon System' in *Understanding the Earth*, ed. I. G. Gass, P. J. Smith & R. C. L. Wilson (Artemis Press, 1972).

Other examples

E. Morgan, *The Descent of Woman* (Corgi, 1973).
S. J. Gould, *Ever Since Darwin* (Penguin, 1980).
S.J. Gould, *The Panda's Thumb* (Norton, 1981).

3

SCIENTIFIC KNOWLEDGE AS POLITICAL AUTHORITY

Jonathan Harwood

What attracts people to study science? Is it perhaps the prospect of consensus? ... that science permits persons from a variety of national, racial or religious backgrounds to reach agreement on the structure of nature? If we consider other forms of knowledge — philosophy, theology, politics, economics — they seem to be just as intellectually challenging as science, but only rarely do experts in these fields achieve the level of consensus characteristic of science.

Liam Hudson's psychological studies of English schoolboys specializing in the arts or sciences are interesting in this regard. Hudson found that these two groups differ as much in personality as they do in verbal or mathematical ability, that personality channels the *ways* in which schoolboys use their abilities. In general young scientists were more interested in impersonal aspects of their surroundings and sought to avoid controversy, especially when it was associated with people. They disliked ambiguity or conflicting ideas and thus were better at solving problems with one right answer than those which were open-ended.

On this basis one might expect many scientists to find politics rather irritating. Certainly a number of spokesmen for British science in the 1930s had grown impatient with political leaders' apparent inability to sort out the economic crisis. Said one:

Scientific research has a fatal fascination. 'Live in the serene peace of the laboratory', said Pasteur. Once a man has tasted of that austere discipline and calm, he will not readily exchange it for the wrangling spirit and the unintelligent obsession with trifles which are all too often found among politicians and the press.

Frequently, it was suggested that politicians could benefit from a more scientific frame of mind.

As much as we might like to turn away from the strife of politics and escape to the laboratory, it is a very dangerous choice. Like it or not, scientific knowledge is used in the political realm in ways which sometimes rebound upon the scientific community, bringing unwelcome interference with research. Science has succeeded admirably in showing how through understanding the potentially destructive forces of nature we can harness them to human purposes. However, unless scientists themselves learn to understand the political impact of their work, and to act accordingly, they will continue to be buffeted helplessly by political winds.

Nature constrains

Everyone recognizes the contribution of scientific inquiry to the advanced technologies so central to our everyday lives. Less well known but just as significant is the role of scientific knowledge as a source of authority in the political realm. In virtually every society politics concerns the clash of sectional interests. Policy-making invariably means treading on some-one *else*'s toes in order to promote what *we* define as just or necessary. Resistance to a political measure must somehow be minimized, if possible by demonstrating to our opponents that their preferences are impractical or immoral. One way to achieve this is by giving our opponents reasons which are so complicated that laymen have no way to challenge our policy. They must then seek experts who can fight this battle for them, but this may be difficult for interest groups lacking financial and other resources. In this respect scientific knowledge is no better than any other esoteric body of knowledge. In another respect, however, science is distinctively powerful. The anthropologist Mary Douglas suggests that in every society four kinds of justification are commonly used to prevent people from doing what they want to do. *Time, Nature, God* and *Money* constitute the universal constraints on freedom. When a child asks for ice cream, we reply: 'We're in a hurry', 'it's bad for your teeth', 'it's wrong to eat between meals', or 'we can't afford it'.

From this point of view scientists can be seen as a secular priesthood who, as specialists on the natural world, are in a position to exert substantial influence upon political debate. From its emergence in the West in the sixteenth and seventeenth centuries, science has always appealed to a wide variety of interest groups as an apparently unimpeachable source of authority. Derived simply from the powers of observation and reasoning possessed by everyone, science's claims appeared rooted in the real world, open to challenge and thus improvable, unlike other forms of knowledge such as art, religion or philosophy. Precisely because it seems so *un*political, science has long been extremely attractive to those engaged in political combat.

For example, in late seventeenth and early eighteenth century England different conceptions of matter were an important political resource. Radical groups like the Diggers, Levellers and Freethinkers drew upon theories which regarded matter (and man) as infused with spirit (or God's forces). Matter was thus seen as active and self-organizing; laws of nature were inherent in matter itself. From this view of matter the radicals argued that human beings possessed wisdom via a direct link with God and did not, therefore, require the Anglican Church as a middle-man. The radicals wanted the political power of church and king replaced by parliament in a new republic. The defenders of the church's central position in the monarchy, on the other hand, drew upon Newton's conception which treated matter as lifeless, brute, passive and stupid. God's will (in the form of the force of gravity, for example) was, therefore, required to bring matter into motion and generate order in nature. By implication, human beings could not organize themselves without the church's intervention and interpretation of God's will.

In the nineteenth century biological thought in general — and evolutionary thought in particular — were frequently drafted in support of all manner of political programmes. Some 'social Darwinists' (as they have sometimes been called) — such as the American industrial magnates Andrew Carnegie, John D. Rockefeller and their admirers — seized upon the idea of within-species competition. They saw fierce economic competition as a progressive force, guaranteeing that firms which were 'fit' would survive and expand while the inefficient and otherwise 'unfit' would be eliminated. Others, such as the Russian anarchist Kropotkin, emphasized the importance of co-operation in securing social progress and also found support in Darwin's work. Similarly, socialists throughout the West drew upon Darwin's discussion of between-species competition in order to affirm their belief in social progress through class struggle. Ernst Haeckel, Darwin's champion in Germany, derived yet another version of social Darwinism which stressed social progress through the struggle among nations and races, a process demanding individuals' self-sacrifice in the interests of the group's survival. With its anti-church, anti-democratic, anti-liberal and anti-working-class message, this political programme was opposed to anarchism, socialism and competitive capitalism and was one of the formative influences upon German fascism.

Such diversity in evolutionary political philosophies was not simply the product of different national contexts. In Britain, for example, Herbert Spencer, while arguing that society was a social organism, insisted upon this analogy's limits. While in an animal only nervous tissue possesses feeling, in society *all* individuals have this capacity. Consequently, while it was right that in the animal all parts were subservient to the nervous system, a society lacks this single overriding consciousness, and thus individual persons' wellbeing should not be subordinated to an overarching body like the state.

Individual freedom was prior; the role of the state was merely to respond to the wishes of its citizens. On the other hand, thinkers from other British political traditions, such as the Fabian socialist Sidney Webb, pushed the idea of society as an organism much further. He emphasized the *integrated* character of biological systems, concluding that Spencer's stress on individual freedom for the parts was just as destructive in societies as it would be in the body of an animal. Thus, Spencer and Webb were each trying, in different ways, to enlist a biological concept for political ends.

Two points need to be made about social Darwinism. First, Darwin was more interested in explaining the distribution of animals and plants in nature than in making a political statement, and in *On the Origin of Species* he said nothing about human beings. He was undoubtedly opposed to most of the political movements which sought authority in his work, but once his theory was published, he had very little control over the uses to which it was put. This does not mean, however, that most of the political extrapolations from his work were 'incorrect'; the theory of natural selection, like any theory, contained a *variety* of concepts, any one of which could become a starting point for further developments (whether scientific or political) which the theory's inventor could not anticipate. Second, though social Darwinism had its heyday in the nineteenth century, it remains a common means of justifying unequal treatment in our own time. In 1979, the principal of a prestigious London college argued that since there were too many universities and polytechnics, some would have to be sacrificed for the sake of the others. When resources were scarce, only the fittest competitors could survive.

In the twentieth century a related form of political morality was extracted from the biological and psychological study of man by 'eugenics movements' in all industrial societies. Until the mid 1920s, the majority of interested American academics took it for granted that most socially significant aspects of human behaviour were genetically determined: e.g., intelligence and other mental abilities, personality, criminality, morality. Genetic 'defectives', it was argued, had to be prevented from reproducing. Thus 'hereditarian' theories of behaviour were invoked in the USA to justify the legalization of (in some cases mandatory) sterilization for prostitutes, the feeble-minded and car thieves in twenty-one states between 1907 and 1928. In Britain the compulsory segregation of mental defectives in asylums was legal from 1913. At the same time it was argued that middle-class professionals, who earned their living through the exercise of superior brain-power, should enjoy tax reductions and family allowances so that they could afford to have more (genetically gifted) children and to educate them. Furthermore, some British eugenists argued, it was quite wrong for Britain to be ruled by a class which was enfeebled by in-breeding and which transferred its wealth to the first-born child rather than to the genetically fittest. Thus were the laws of heredity seen to necessitate the redistribution of political power and social advantage.

In the United States a decisive factor in the passage of a discriminatory immigration law in 1924 had been the evidence from intelligence testing in World War One which purported to show that much higher proportions of (recent) immigrant populations from southern and eastern Europe were intellectually sub-standard than among (longer established) immigrant groups from Britain and Scandinavia.

In educational policy a new, statistically-based psychology became influential. By the early twentieth century the channelling of working- and middle-class children into vocationally and academically oriented secondary schools, respectively, often on the basis of teachers' recommendations, had come under attack in the USA on the grounds of its apparent political bias. By the early 1930s, however, selective schooling had acquired a scientific rationale; performance on intelligence tests was said to be largely genetically determined and thus, it was claimed, provided a socially fair criterion for selection. One survey in 1932 showed that three-quarters of a sample of American urban education districts were using such tests to

allocate pupils to schools. The proportion using tests at that
time in Britain was substantially smaller, but in the influential
Hadow Committee's Report of 1926 (recommending free
secondary schooling for all children in schools deemed 'appropriate' to the child's tested capacity at age eleven) as in arguments for 'streaming' or 'tracking', the idea of the inherited
character of mental capacity was strongly represented by leading British psychologists like Charles Spearman and Cyril Burt.
The testing of innate mental ability was often championed by
educational reformers who saw it as a way to identify bright
working-class children whose talent would otherwise have
been lost to the nation.

After World War Two it was no longer so widely agreed
that intelligence tests were fair to working-class and ethnic
minority children. In the USA, Alison Davis began in the
late 1940s to argue the class bias of testing, and in Britain
during the 1950s environmental influences upon test performance were being emphasized by critics of educational selection.
If intelligence was substantially moulded by home and class
background, testing could not be relied upon to spot born
'winners'; unselective (or 'comprehensive') schools were
advocated as a better way of realising pupils' potential. Thus,
although educational reformers had abandoned testing, they
continued to invoke psychological and sociological research in
support of their cause.

These are but a few examples of the political significance
ascribed to particular areas of scientific knowledge in the
modern era. But what, you may object, is wrong with building policies upon scientific foundations? Surely any realistic
policy must be consistent with our understanding of relevant
areas of nature if it is to succeed? This is perfectly true,
of course; high-rise housing, for example, is wasteful if it
clashes with basic human psychological needs. What is crucial,
however, about the deployment of scientific knowledge
in political debate is that scientists cannot control this process.
First, a given body of fact does not logically entail only *one*
political conclusion. This point has often been summarized

by philosophers as the impossibility of deriving an 'ought' from an 'is'. If intelligence differences were agreed to be entirely genetic, it would not necessarily follow that schools should be specialized or classes streamed. We might decide that the social advantage of mixed-ability classes outweighed the losses in teaching efficiency. Conversely if intelligence differences were entirely environmental in origin, it would not follow that selection or streaming are inappropriate. Though the state of scientific knowledge will constrain the available policy options, it cannot tell us which of them is best.

Second, the deployment of science is usually *selective*. This sometimes occurs through giving differential publicity to available facts. When Richard Nixon's expert Presidential Commissions, looking into the possible dangers of cannabis and pornography, recommended liberalizing American legislation in these areas, he chose to ignore their findings. Similarly when James Coleman's report in 1966 on the extent of educational opportunity in the USA came to conclusions at odds with government policy, the report was shielded from public attention and never acquired much influence upon American educational policy. Scientists sometimes take satisfaction from requests by policy-makers for further research into policy-relevant areas of science; it looks, perhaps, as though knowledge is being accorded a key role in decision-making. Faced with the prospect of additional research funds, however, it is all too easy for scientists to disregard the fact that policy-makers are thereby able to buy time, hoping that in the meanwhile a dispute will cool off and opposition will lose momentum.

Another way in which knowledge is used selectively arises when the area of expertise employed is controversial. This is common in areas of policy-related science and has been evident recently in debates over issues such as the heritability of intelligence, the safety of nuclear power and pesticides, or the causes of lung cancer. Where experts disagree, science cannot be decisive, and interest groups must pick their experts very carefully (as we will see below). In short, when science enters the political arena, it usually does so on politicians' terms.

That scientists are 'on tap but not on top', however, does not mean merely that they cannot control the uses to which their expertise is put. Much more fundamentally it means that the organization and practice of scientific research itself may be altered by constant demands from patrons in the wider society for scientific testimony. For as scientific knowledge is drawn upon as a source of authority, the scientific community itself becomes a potential target for political attack. Often such backlash seems to catch scientists off guard; because their own motives in conducting research were disinterested, they are surprised and annoyed to see how differently their knowledge is received in the world of politics. In the pages below, both the appropriation of science as authority and the political consequences of this for science are illustrated by reference to recent welfare policy in the United States.

The impact of political visibility upon research

By the late 1950s the rapid pace of post-war automation plus job discrimination had produced high levels of unemployment and juvenile delinquency among blacks in the northern cities. In response, the Carnegie and Ford Foundations began to sponsor research on this 'social dynamite' as well as pilot programmes designed to raise morale in decaying inner-city communities. By the early 1960s, the Civil Rights movement, organized to fight for basic political rights for southern blacks, was a substantial force whose political ambitions had begun to broaden to include discrimination in the north. When Lyndon Johnson assumed the presidency late in 1963, therefore, his advisors were keen on a package of welfare programmes, patterned on those of the Ford Foundation, which aimed to extend equal education and occupational opportunity to the (disproportionately black) urban poor. Made law in 1964, this broad array of 'War on Poverty' programmes emphasized education in ways that may be familiar to British readers, especially those who recall the Community Development and Educational Priority Area programmes of the late 1960s and early 1970s: job-training for the 'hard-core unemployed' and enriched nursery schooling ('Project Head Start') to compensate for the environmental deprivation of children from poor families.

A central assumption of the War on Poverty was that poverty could be cured through altering behaviour: through education, people of low ability and limited skill could acquire more demanding and desirable jobs. Ignorance, the President argued, was the 'taproot' of poverty. Accordingly, one of the main aims of the compensatory education programme was to raise the tested intelligence of its children so as to help them cope more successfully with the school system, then with the job market. Various pamphlets issued in the mid-1960s by the Office of Education and Department of Labour, therefore, declared that innate differences in intelligence between races did not exist. The Governor of California's Commission on the Los Angeles riots concluded similarly that slum children's academic failure (a source of frustration thought to be partly responsible for rioting) was rooted in their earlier childhood experiences. Environment, it appeared, was all-powerful, and a panel of academic experts brought together by the White House to examine the feasibility of enriched nursery education largely confirmed this view.

Here, then, was a political programme which sought justification in scientific expertise. The experts consulted, however, did not represent the entire range of opinion within their disciplines, a fact which was later to prove embarrassing. Moreover, the existing research on compensatory educational methods had concentrated primarily on setting up and publicizing working programmes which produced immediate increases in children's educational performance. Such projects had not yet been subjected to critical evaluation and long-term testing, and some of the academics then active in this research have since conceded that they lacked an adequate theoretical basis for Project Head Start.

Why were these deficiencies overlooked by most scientists in the relevant areas at that time? The answer emerges from looking at the impact of the War on Poverty upon research related to pre-school education. The experts appear to have been extremely enthusiastic at this time about the potential social contribution of their knowledge. Foundations and the federal government gave financial support to their research on an unprecedented scale. Instant popularity, however, is not conducive to careful reflection upon the solidity of one's own enterprise, and more than one academic became preoccupied with grant-hunting and bandwagon-hopping. Believing in the promise – if not the actual accomplishments – of their work, the experts could hardly afford not to collaborate with federal policy-makers in designing and implementing programmes which seemed to be on the forefront of social reform and to inaugurate a new era of public support and recognition for their areas of expertise.

But what, exactly, was the significance of scientific knowledge for War on Poverty policy? The scientists themselves sometimes expressed the view in the mid 1960s that post-war research in their discipline had prompted the new reforms: before World War Two such educational policies had been unthinkable because it was believed that intelligence differences were largely innate and thus unalterable. As with so many self-serving ideas, this too was self-deluding. A variety of analysts of the War on Poverty have concluded that the social science literature consulted in the construction of the War on Poverty as a whole was used merely to provide window-dressing, not a foundation. When we look more specifically at Project Head Start a similar picture emerges. The basic idea of enriched nursery education was conceived by policy-makers *before* a panel of experts was appointed to report on its feasibility. Though the panel recommended thorough and intensive compensatory education, initially on an experimental basis, for relatively few children in the first year of the programme, government planners — responding to Head Start's enormous popularity — decided to implement far more limited programmes for many more children. This looked like a better way to catch the public eye and help the government to secure greater sums from Congress for subsequent years' spending. Faced with the problem of selling a political package, the planners had little interest in checking out the reliability of the expertise they had selected. Scientific knowledge was thus imported, not in order to explore possible policy implications, but in order to justify politically expedient policy.

Once the War on Poverty was under way, the costs of academics' involvement in policy soon became apparent. Subject to constant pressure to design programmes and produce successes very quickly, several of the most eminent psychologists involved protested that good work could not be done in an atmosphere of continuing crisis and urgency. The breathing space which they sought never came. The race riots of 1964–8 encouraged criticism of the War on Poverty, and growing US involvement in Vietnam competed with the War on Poverty for financial resources. At the end of the 1960s, as the War on Poverty was winding down through dwindling political support, a new group of academic experts moved into the limelight, citing the failure of many Head Start programmes to produce lasting gains in intelligence test performance. The explanation (argued A. R. Jensen and Richard Herrnstein in the USA and H. J. Eysenck in Britain) was that the scientific rationale for compensatory education was false: intelligence differences were largely due to heredity rather than to environment and thus could not be readily erased.

This provoked a bitter academic controversy, continuing into the mid 1970s. Several of the scientists who had devoted their expertise to the success of Head Start now became Jensen's major critics. Such controversy can hardly be ascribed to the cantankerous personalities of those involved; academics who had collaborated with Jensen and admired his work prior to the dispute thereafter became hostile. Nor can the controversy be explained by the intrinsically incompatible character of 'hereditarian' versus 'environmentalist' theories of behaviour. Contradictory though they may be, such theories had managed to coexist relatively peacefully throughout most of the 1960s.

To account for the controversy we must recognize that Jensen and his supporters represented a serious threat to the vested professional interests of those experts who were closely associated with compensatory education and had been enjoying government patronage. Jensen put arguments into the mouths of the political enemies of Head Start; reduced federal budgets for the War on Poverty meant that pre-school-educational research begun in the 1960s could no longer be so freely pursued in the 1970s. Anyone with an axe to grind against the social sciences was now sitting pretty: the practitioners of these 'pseudo-sciences' could not agree among themselves, and those of their theories which were put to the practical test did not work. Having entered the political arena in the early 1960s with great expectations for their disciplines' growth as well as their contributions to social reform, these social scientists now found themselves highly vulnerable to changing political trends which they had not anticipated.

It is evident that scientists cannot hide from the constant search for political authority which goes on around them. Scientists will have to come to terms with the fact that their knowledge is likely to remain an attractive source of legitimation for political movements for a long time to come. Furthermore, there is no way such movements' interest in science can be avoided. Since the scientific community lacks independent financial resources, it may be possible for scientists doing small-scale or paper-and-pencil research to shop around for the best deal from a sponsor, but expensive and technically sophisticated disciplines will necessarily require the high level of financial support which only government and industry can offer.

What then can scientists do? In the years of plentiful government funding since World War Two the scientific community has perhaps come to take its own importance too much for granted, has been able to retreat to the laboratory without having to worry much about its public image. This golden age is rapidly disappearing. In a time of economic recession priorities for government and industrial expenditure become the subject of prolonged argument: has our spending on science paid off as we were led to expect? Should we intervene more directly in order to ensure that our money is spent properly? Unless scientists take account of the impact which their work is making upon powerful groups in the wider-society, seeking if necessary to dissociate themselves from what is done in the name of science, the scientific community can easily become the target, perhaps quite unjustly, for damaging attacks. One might consider, for example, whether some of the difficulties which evolutionary biologists now face in confronting 'scientific creationism' (discussed by Edward Yoxen in Chapter 4) could have been forestalled had scientists taken time to think about how evolution is perceived by certain groups, instead of assuming that biological theory was an unalloyed good — the more, the better.

Despite all their inclinations to the contrary, if scientists wish to preserve that 'austere discipline and calm' of the laboratory, they will have to learn to engage where necessary in the 'wrangling spirit' of politics. The history of the scientific community's dealings with patrons can be instructive in this respect. If scientists spent a bit more time studying the successes and failures of their past, they might be better equipped to deal with their future.

Questions for discussion

1.   If one cannot logically derive a unique political conclusion from a scientific fact, why do political figures continue to invoke scientific authority?

2. Is it correct to regard the scientific community as relatively powerless compared with its patrons? Aren't the latter dependent upon scientists' co-operation?

3. It is perhaps not very surprising that the *human* sciences should prove so readily politicized. Can you imagine how other sciences might be similarly incorporated into political debate and become subject to the kinds of pressures described above? (e.g., geology, toxicology, radiobiology, chemical or nuclear engineering, etc.)

Further reading

Richard Hofstadter, *Social Darwinism in American Thought* (Beacon Press, 1955).

The author analyses the great variety of ways in which Darwinism was extrapolated to society up to World War One and suggests why these interpretations shifted during the late nineteenth century.

Clarence Karier, 'Testing for order and control in the corporate liberal state' in N. Block and Gerald Dworkin (eds.), *The IQ Controversy* (Quartet Books, 1977 and Pantheon Books, 1976).

The author discusses the connections between intelligence testing, eugenics and immigration in early twentieth century USA.

Peter Marris and Martin Rein, *Dilemmas of Social Reform* (Routledge and Kegan Paul, 1972).

The authors present a case study of the formulation of anti-poverty policy and the way in which it drew upon social scientific expertise.

Carol Smith Gruber, *Mars and Minerva: World War One and the Uses of the Higher Learning in America* (Louisiana State University Press, 1975).

The author demonstrates the unfortunate consequences of American academics' zeal in putting their expertise at the service of the war effort.

Brian Simon, 'Classification and streaming: a study of

grouping in English schools, 1860–1960' in his *Intelligence, Psychology and Education* (Lawrence and Wishart, 1971 and Beekman Publications).

The author discusses the role of intelligence testing in the evolution of inter-war education policy leading up to the 1944 Education Act.

4

DARWIN AND/OR DOGMA?

Edward Yoxen

It is a shock to realize that about one-sixth of the population of the richest and most powerful nation on earth believe themselves literally to be descended from Adam and Eve. It is worrying to think that among this group millions of Americans deliberately and consistently reject the Darwinian theory of evolution, that they are uninterested in what scientists might say about scientific theories, that they are totally unimpressed by the millions of hours researchers have spent gathering data and building theoretical models and that their tolerance is so limited that they are far from happy with the continued propagation of Darwinism as an alternative to their system. These, then, are extreme attitudes, also to be found in Britain and other European countries, but given particular importance by the shift to the right in America that brought Ronald Reagan to the White House. Creationism, the view that the Genesis account of God's creation of the world and all living things is the *literal* truth, is now part of the intellectual and moral landscape, particularly in the USA. The creationist movement offers an explicit threat not only to the practice of research but also to the moral and political traditions of our contemporary scientific culture. Creationists seek to put the Bible back into politics and to take science out of people's consciences.

From one point of view the recent controversies, which are described below, are yet another phase in the immensely complex working out of the implications of Darwinian evolutionary theory, a process that began, say, in the 1850s.

If species are formed over long periods of time by evolution, and not created by God's *fiat* in an instant, as creationists claim; if all organisms fit into a pattern of common descent, a tree of organic forms with its roots in the pre-biotic soup and human beings at the top of its highest branch; if natural selection through the more prolific reproduction of better adapted organisms is the operative principle behind this changing system, then what makes human beings different or special? If we are zoologically kin to monkeys, what does this say about our morals? From this perspective what do terms like individual responsibility, reason, spirituality, faith, commitment, altruism and political consciousness mean? Issues such as these are central to any world view and the rise of evolutionary theory in the mid nineteenth century forced their reconsideration. It is not surprising that they continue to be debated, as major political upheavals place intellectual and moral systems in jeopardy. It is the terms in which the debate is now being carried on and the political significance of a certain style of argument that is so disturbing.

Darwinism is, in part, about the nature of man. In its contemporary neo-Darwinian form it is also a unifying theory of enormous scope. It seeks to embrace every species and every organism that ever has existed, exists now or will exist. Moreover it seeks to relate many phenomena of anatomy, physiology, heredity and ecology to the evolutionary process. The opposable thumb, the Krebs cycle, the universality of the genetic code, the single-egg reproductive strategy and the symbiosis of leguminous plants and nitrogen-fixing bacteria are, we are taught, the products of evolution. Yet this vast intellectual system rests on rather meagre foundations of evidence. Whilst it is possible to provide countless *illustrations* of the basic ideas, it is much harder to provide examples where neo-Darwinian evolutionary theory is actually *tested*, not least because many of the events under scrutiny occurred millions of years ago.

Darwinian evolutionary theory can be subdivided into two propositions; firstly the idea of evolution and common descent, that is to say that similar organisms separated in time can be grouped together in an evolutionary schema, and secondly the idea of natural selection. Charles Darwin argued in his *On the Origin of Species* for the existence of a struggle in nature which would force the elimination of less competent or fit individuals. Variations in structure, physiology and behaviour are, the argument runs, continually tested as possible improvements in the adaptation of particular organisms to their environment, in competition or interaction with other individuals and species. Natural selection is expressed through differential survival of different types and, over long periods of time in which major environmental changes tend to occur, new species evolve. In Darwin's view, God created the simplest organisms in the remote ages of geological time, an idea evolutionary biologists now reject in favour of various chemical or thermodynamic theories of the origin of even the simplest and earliest organisms. With the rise of Mendelian genetics in the twentieth century we now have an account, unavailable to Darwin, of how hereditary characteristics are transmitted between generations. We can now see how variations can be conserved. With the creation of population genetics in the 1920s and 1930s came a recasting of evolutionary theory, since the differential survival of particular types of individual is a matter of interactions or statistical processes amongst a population of organisms, and this picture of evolution was renamed neo-Darwinism.

Interestingly, the study of evolutionary processes has become an area of excitement and theoretical controversy among biologists in the last ten years. For example, postulated links in the evolutionary pattern depend on judgements of similarity and difference, i.e., on classification. Recently biologists have begun to question the basis of conventional categories used in classification. Some query the idea of species; others argue that the idea of common descent is an over-interpretation of the available information or even inessential. Another controversy concerns the increasing evidence that speciation, the formation of new species, can occur through discontinuous change rather than through the continuous accumulation of minor variations. Another argument surrounds the role of natural selection. Some population biologists have been arguing for a while that random statistical processes are sufficient to explain the predominance of particular characteristics in populations, without recourse necessarily to the idea of selection. Also some highly controversial data have been published recently that are held to show that, contrary to Darwinian teachings, acquired characteristics are inherited. Finally the application of Darwinian notions of adaptation to human evolution has proved highly contentious. Some biologists have tried to link together the mode of reproduction/sex ratio/behaviour and group structure of particular species under the banner of adaptation, for example in studies of ants and bees. Some have tried to extend this 'sociobiology' to human communities and have been condemned for simplifying and misrepresenting human nature and the causes of social order.

All these areas of disagreement can be viewed as signs of weakness and confusion or of inherent strength through endless renewal. For scientists they are, or should be, the latter. By creationists, however, they are often regarded as welcome signs of uncertainty and doubt, since evolutionary theory is vilified as the cause and expression of moral decay, an instance of and contributing factor to the weakening of religious faith, the erosion of traditional moral values and the spread of permissive or decadent social practices. Perhaps more accurately one should say that some scientists are worried that theoretical ferment will be exploited by outsiders deeply hostile to science, since my impression is that for the foot-soldiers of the creationist army — and the military metaphor is not inappropriate — such niceties as the fact that all theoretical knowledge is periodically rebuilt are irrelevant to their desire to purify social attitudes and beliefs from the pollution of scientific humanism. For them the old simple truths have been pushed aside by the godless and erroneous teachings of degenerate egghead liberals. Their goal is the suppression of evolutionary thought. For the moment in America the ring is being held by the courts and an interpretation of the constitution which seeks to guarantee freedom of religious belief.

Creationism as controversy has a long history reaching back into the early nineteenth century, when its position as an orthodoxy was first threatened by emergent evolutionary ideas. But it is only the events of the 1960s and 1970s that will be discussed here. I shall suggest that in this period the structure of the controversy between creationists and biologists has turned around what the term 'theory' is held to mean. I shall also suggest that the present resurgence of creationism is a resolution of political forces specific to the United States in a period of recession, although similar, less dramatic events are occurring in Britain. This is its real significance. Creationism is not only about how to teach biology: it is also an illustration of how certain groups in society seek to suppress critical, inquiring habits of thought, to outlaw certain forms of knowledge and to reaffirm highly conservative economic and moral notions in a period of crisis. First, let us consider the term 'theory' in science.

Theories as fallible knowledge

Theories are sets of general statements about some restricted part of the natural world and some of the processes occurring there. These statements may be logically related through deductive steps or they may be more loosely linked together. They may be statements of great generality, thought to apply throughout space and time, or they may be much more restricted in range, built up cumulatively or inductively from a series of experiments from which a model emerges. They provide, indeed they force on us, particular ways of describing events. They allow us to relate causally particular phenomena to others: that is, to see phenomenon A as caused by phenomenon B, or to see B as providing an explanation for the occurrence of A, and only A. They marshal evidence, and they allow predictions to be made, for example, that if *B* occurs again, then *A* will also occur again.

To take an example from the evolutionary field, scientists have related changes in the coloration of the wings of the English peppered moth to changes in the colour of exposed surfaces due to industrial pollution, through the theory of evolution. The newer dark-winged moths tend, it is said, to survive more easily on blackened surfaces and outbreed their light-winged relatives. One can predict that as pollution control changes buildings and foliage to their pristine state, the situation will reverse; and indeed this seems to be the case. Note that if we didn't find this latter prediction borne out, we would have several choices. One would be to say that the theory of evolution was quite wrong because it had failed this test. Another would be to say that there were other factors operating, which were not inconsistent with the theory but just unknown to us at present. Or we might say that we just didn't believe the results or want to think about them at the moment. At different moments, scientists respond in each of these ways. Sometimes theories are indeed cast aside forever; more often they are refined and rebuilt; and sometimes the problems are pushed aside and ignored, because in other areas, seen as more important, the theory works very well. Note also what a creationist might say to the peppered moth data. One response would be to say that what is noted here is a trivial, but not uninteresting change, which says nothing about the origin of species, which are, (according to the 'scientific' creationist view) created by God with a certain limited variability, and that the results do nothing to test or to confirm an erroneous theory, namely evolution by natural selection. They simply prop up a misconception.

There are two important conclusions that we should extract from the argument so far. Firstly, theoretical knowledge is provisional and fallible. If science means anything, it therefore means the possibility of change and development, as all scientific knowledge is theoretical. At the same time, it is also clear that theories can be accepted uncritically, promoted dogmatically and retained irrationally even when in difficulties. They always have their limitations, even if scientists choose or are taught to ignore them. Secondly, given the possibility of alternatives implied by the idea of fallibility, it is important to have some way of putting rival theories or even a theory without a visible replacement, to the test.

In practice, as I have tried to show above, it may be difficult to set up a meaningful test and there are real problems in deciding how to respond rationally to its results. If this is so then how can you evaluate the standing of different theories? Testability and the possibility of replacement seems to be an important notion to apply if possible. Some philosophers have tried to use predictive power as a criterion; others say that scientists operate in terms of group interests and values, that they stick with what they understand and know well, unless there are powerful reasons for switching horses. Now let us go back to creationism.

Evolutionary theory as dogma

At the beginning of the nineteenth century, an important tradition in British thought was natural theology, which sought to prove the existence of God by examining the fabric of nature. Such intricacy, such perfect fittedness for use, such beauty in living things surely implied the existence of a designer. Darwin himself was educated in this tradition, although his work, published in 1859, seemed to some to contradict it utterly. This is indeed the essence of the creationist objection, that evolution by natural selection seems to remove God's designing presence from nature. This, they suggest, cannot be; for complex organisms and ecosystems cannot be simply the result of chance operating within the framework of a self-contained, evolving natural world. To this biologists can reply, if they wish, that God designed the system by which nature designs itself. Darwin himself felt that his theory was no threat to arguments for the existence of God. The doctrine of special creation – the Genesis view which natural selection replaced – had simply misrepresented the relation between God and nature. After some trenchant debate in the 1860s, orthodox theology in Britain and elsewhere took on board the Darwinian theory. But occasionally there are periods of powerful religious and cultural reaction; and the 'old' argument from design is born again.

In 1925 the Tennessee legislature passed the Butler Act, forbidding the teaching of evolution in the public schools of that state. In May of that year, John Thomas Scopes, a science teacher in Dayton, consented to be the defendant in a court test of the law. He stood trial in July 1925. Defended by the American Civil Liberties Union, Scopes was represented by the noted barrister, Clarence Darrow, at whose direction he pleaded guilty. In the event the prosecuting counsel were more embarrassed than vindicated by the outcome of the trial, but the Act remained in force. In 1926 and 1927 similar laws were passed by the legislatures of Mississippi and Arkansas; the latter statute remained unrepealed, although unenforced, until 1968, when the US Supreme Court declared such laws unconstitutional.

The Scopes trial can now be seen as a struggle between progressive and secular thought, based in the northern industrial cities, and the moral and political values of country people in the southern states, who felt threatened by industrialization, inflation, a crisis in agriculture and the encroachment of city life. Certainly, the teaching of evolution in US schools was powerfully inhibited by events of this kind, despite the efforts of modernizing biologists to revise the school curriculum. In the 1950s, as part of a more general reform of US high school education, an important pressure group and educational resource centre was established in Colorado to produce new biology textbooks for mass sale by commercial publishers. The Biological Sciences Curriculum Study programme set out deliberately to introduce evolutionary thinking into the American school systems, as they saw it, to make up the lost ground. In this respect they were outstandingly successful. However, their efforts also stimulated the formation of a number of pressure groups who attacked the new biology courses, lobbied for the removal of the new teaching material and for the teaching of creationist theories alongside neo-Darwinian accounts. Some of these pressure groups, amongst them the Creation Research Society and the Institute for Creation Research, recruited only highly trained scientists and engineers and made much of this in their publicity. In California they enjoyed the sympathy of the conservative Superintendent of Education, Max Rafferty. Their demand was for 'equal time' for creationism, and by implication for equal status for creationism as a scientific theory. This goal they failed to achieve in the early 1970s although in some states only by the narrowest of margins, but they managed to force some major revisions in all teaching texts, to get some publishers to delete virtually all references to Darwin and Darwinism from their books, and to secure a place for creationism in the school curriculum, though not necessarily in biology. It is fascinating to see some of the changes wrought in textbook formulations by creationist lobbying, which were intended either to reduce the dogmatism with which evolution was presented or, going somewhat further, to decrease its status as a theory.

The disputes over evolution in parent-teacher associations, in school boards, in state legislatures and in the courts were often bitter. But at least as much rancour was stirred up by a related educational development promoted with federal funds, *Man: a Course of Study* (MACOS), which was a high school course, based on animal behaviour and the anthropological study of a Canadian Eskimo tribe that practised infanticide and open marriage. Its educational rationale seems to have been to get students to think about the basis of morality, and social organization, and to confront head-on the fact that moral conventions and social institutions do vary widely in human communities. The course soon drew harsh criticism as subversive, offensive and dangerous. In West Virginia the campaign to block its use in high schools led to pickets, arson and a bomb attack on a school. Depressingly, even the idea that students might think comparatively about and then re-affirm existing moral standards from a broader perspective was anathema to the critics. For them it was not the job of schools to develop the capacity of moral reasoning. Fundamentalist religion is linked to private socialization, in the home. The creationists claimed that federal funds were being used to promote a covert secular religion in the guise of biology, and citing the constitution in their support, they claimed that the freedom of belief guaranteed to their children was being denied in the schools. In some part this view was upheld in the courts and some provision has been made for the teaching of creationist ideas.

This struggle over the nature and purposes of education is still going on. An 'equal time' bill was just defeated in Georgia in 1979, having been energetically promoted by members of the Southern Baptist Convention, the largest Protestant sect in the United States. In 1981 a California judge ruled that the teaching of evolution in schools does not necessarily under-mine the religious beliefs of those who accept the biblical description of human origins. This claim had been made by Mr Kelly Segraves, director of the Creation Science Research Center in San Diego, on behalf of his son. The judgment effectively reaffirmed the compromise erected in 1975 re-quiring the moderation of evolutionary dogmatism and its representation as theory. Whether that compromise will hold remains to be seen. In late 1981 the Arkansas Federal Court considered and rejected as unconstitutional a 'balanced treatment of creation—science and evolution—science act'.

This moral backlash is connected indirectly with the difficulties that have beset the US economy in the last few years — with declining competitiveness in home and inter-national markets, serious loss of profitability in major indus-tries like steel and automobiles, rising unemployment and rising inflation. Reagan's election programme laid heavy emphasis on the regeneration of the economy by stimulating the supply of new products and technologies, rather than by

the government purchasing and changes in taxation to stimulate consumption. Let corporations and entrepreneurs make a
healthy profit again, by cutting taxes, removing regulations,
and reducing the price of labour, and the cycle of wealth
creation will pick up again. These economic themes are familiar in Britain. With this economic programme comes also a
moral crusade, which combines a re-assertion of self-help,
a stress on 'traditional morality', and a visceral belief in the
goodness of the American way. Again, these themes will be
familiar in Britain. But Reagan's election was bound up with
a political regrouping of forces, in which religious groups
played a vital role.

The mid 1970s saw changes in attitude and strategy
amongst pressure groups on the extreme right of American
politics. Outraged by the confusion and apparent weakness
of the Carter presidency and dismayed by the thought that
the Republican Party might again choose an unsatisfactory
presidential candidate, various influential 'political mechanics'
began to build up a coalition among single-issue lobbies,
that were concerned, for example, with stopping gun control,
blocking equal rights for women, or abortion. In this coalition
various fundamentalist church leaders, some with their own
TV stations and multi-million dollar fund-raising enterprises
behind them, played an important role. One of them coined
the term 'the Moral Majority', to market the idea of conservative opinion, out there in the suburbs, but held back until now
from the decision-making process. With a stress on the value of
traditional family life, on self-help and on fundamentalist
Christian doctrine came also supply-side economics, criticism
of government 'interference' in business and a highly aggressive
foreign policy.

Now the creationist movement is not the same as the Moral
Majority: and the shift to the right in America (and Britain,
Australia and Sweden) is not only a shift in moral values.
But the movements belong together, and reinforce each other.
One of the disturbing features of the operation of the New
Right is the highly organized and successful attack on liberal

politicians as 'liberals'. This is not just a matter of public
school slang, as Thatcherite politicians mock opponents
by calling them 'wet'. It is a multi-million dollar mailing
and broadcasting exercise to mobilize right-wing votes and
to discredit particular politicians as too soft, too 'concerned'
and too broad-minded. The moral tone of this political rhetoric goes far beyond the supposed resurgent decency and lack
of pretension of Carter's 'born again' Christianity to something
harsher and more aggressive; not simply the reaffirmation of
Christian values after the decadent sixties but a return to
full-blooded competition, to social inequality as a virtue
and international strength through overt aggression, all sanctioned by direct references to the Bible. (In Britain its nearest
equivalent in tone and reference is probably the rhetoric of
Rev. Ian Paisley.) Much of the writing of the New Right and
the preaching of its religious wing returns to the theme of
betrayal by those in government by the hopes and desires of
right-wing voters. Too often, they imply, truly radical policies
have been tempered by vacillating politicians. The political
and economic situation is now such that real determination
is the only answer. One of the architects of the New Right
has written of one of his close collaborators.

Paul likes to argue that we are at war. 'It may not be with
bullets,' he concedes, 'and it may not be with rockets and
missiles, but it is a war nevertheless. It is a war of ideology,
it's a war of ideas, it's a war about our way of life and it has
to be fought with the same intensity, I think, and dedication
as you would fight a shooting war.'

The war, let me just reiterate, is against liberalism as a form
of decadence in the United States – or anywhere else.

Against this background what does the creationist upsurge
mean? Firstly, some of the centres of creationist activity
coincide with areas like Southern California and the South-
western states, where high technology industries have been
established in recent years and where older industries have
been relocated because of tax incentives and the cheapness
and relative docility of labour. Creationism is, in part, the

moral panic of a newly wealthy middle class of engineers and technologists in such states glimpsing the prospect of recession. Secondly, despite the attempts to develop a 'scientific creationism' that could be judged against evolutionary theory, the fundamentalist critique is, I believe, anti-theoretical but not anti-science. Truth, for them, is not refined and developed; it emerges from revelation, in one piece, and is passed on authoritiatively without ever needing to be recast. Scientific knowledge simply means systematic, practically tested knowledge, a few basic ideas from which anything can be worked out. 'Theory', in their view, is weak, as yet dubious knowledge as opposed to fact. It is like the engineer's view of science, never speculative, never uncertain and never replaced. Just as the Bible is literally true and needs no fancy interpretation, so science is a collection of equations which give you the answers.

Finally, one cannot but be struck by the explicit conformism of the fundamentalist movement. Even having to justify one's beliefs and standards is seen as a threat. Leaflets for parents counsel withdrawing children from classes where they are asked for opinions or to reason for themselves. Values, it seems, must be inculcated silently and produced on reflex without the mediation of the intellect. This too seems to fit the mental posture preferred by the New Right. Sentiments are mobilized and presidents elected; but the general structures of pluralist democracy within which differences of view emerge and are tolerated are, one feels, a nuisance and a relic. Politics for them is conducted through mass mailings to preselected voters, the private TV station and the private committees that orchestrate national coalition politics. Freedom of thought is a threat to the manageability of the political system, a sign of unpurged weakness, and a chink in the national armour through which tolerance, breadth of vision, humanity, pluralism and individuality might slither in.

Questions for discussion

1. Why do you believe in evolution?

2. How much biology teaching do you think could go on without needing to make any reference to a theory of evolution? What kinds of questions could we not ask? How soon would this be a practical problem?

3. Do you think that regarding the book of Genesis as literally true historically impoverishes our understanding of how myths are created or how people seek metaphorically to express spiritual truths?

4. Do you think it is dangerous to get into situations where a teacher asks you a question beginning 'Do you think ...?'

5. If neo-Darwinian evolutionary theory is roughly correct as a description of human origins, on what basis do ethical systems rest?

Further reading

Dorothy Nelkin, *Science Textbook Controversies and the Politics of Equal Time* (MIT Press, 1977).

A thorough survey of the creationists' activities up to 1976; readable, sociologically informed, usefully interpretative but not polemical.

David Dickson, 'Let there be light!' *Nature*, 284 (17 April 1980), 588–9 (in most large public libraries).

Science writer's view of recent developments; bills, court hearings, books.

Awake! Accidents of evolution or acts of creation? Watch Tower Bible and Tract Society The Ridgeway London NW7 1RN September 1981.

Jehovah's witness view of creation, putting the 'old' argument for design: in my view this is outwardly plausible, but deeply misleading.

John Durant, 'In the beginning there was Darwin', *The Guardian* (16 October, 1980), 13.

Extract from his Darwin lecture to the British Association

for the Advancement of Science in 1980; useful survey, stressing the limitations of neo-Darwinism, although committed to it: argues that creationism shows that values are implicit in *all* science.

D. R. Oldroyd, *Darwinian Impacts* (Open University Press, 1980).

History of science up to twentieth century: clear, readable but not elementary.

Michael Ruse, 'A philosopher at the monkey trial', *New Scientist*, 93 (4 February 1982), 317–9.

An interesting and candid account of what it was like to sit in the witness box in Arkansas in January 1982 and attempt to defend neo-Darwinism as a scientific theory on philosophical grounds.

5

CONTROLLING TECHNOLOGICAL RISKS: THE CASE OF CARCINOGENIC CHEMICALS

Alan Irwin

There can be little doubt that, for many of us, cancer has become an especially dread disease. In western nations such as Britain or the United States, perhaps one person in four will contract cancer and one in five or six will actually die from it. In 1979, cancer killed 120,000 people in Britain. Whilst vast sums have been spent on cancer research (in the USA alone, more than one billion dollars per year) no all-purpose 'cure' has yet been found.

Although connections between cancer and certain occupations have been suspected for centuries, in recent decades awareness has increased that a substantial proportion of cancers may be linked to certain cancer-causing chemical or radioactive agents (known as carcinogens). Evidence has accumulated which identifies carcinogenic agents not just in the workplace (although occupationally-linked substances such as asbestos, vinyl chloride monomer and beta-naphthylamine have received particular attention) but also in a variety of other contexts. Carcinogens in minute quantities have been reported in food, drinks, food additives, medicines, the air, the soil, the water and certain consumer products. The emergence of such data has led to an acceptance that carcinogenic risks are not just restricted to particular occupational groups (although some of these may still be at a greater than average risk) but threaten the population as a whole. One outcome of this has been for certain sections of the mass media to seize on every 'scientific report' which suggests a link between cancer and everyday substances such as caffeine, spinach, weedkillers, beer or cleaning fluids.

The intention of this chapter is not to provide yet another account of the likely carcinogenic risks associated with everyday products, but rather to direct attention towards one important aspect of this issue. Given the technical difficulties and not inconsiderable economic costs which currently surround such matters as the identification and control of suspected carcinogens, how is society to make sound decisions regarding the technological risks to which we are all exposed? The example of British regulatory procedures for the control of carcinogens has been chosen to illustrate this wider problem.

One important dimension of our general question will concern the role of technical experts in such decision-making and the value of the advice which they offer for policy resolution. We will also consider the role of advisory committees and the various kinds of 'pressure group' which attempt to influence overall policies and specific decisions. Although the emphasis will be on the control of carcinogens in Britain, brief reference will also be made both to United States and European regulatory activities. The material in this chapter will provide the basis for a discussion of public participation in technical decision making of this sort.

Carcinogenesis and technical expertise

Before beginning to examine British regulatory structures, it is important for the reader to understand some of the major technical difficulties involved in identifying cancer-causing chemicals and assessing the potency of their effects. The first such factor is **the complexity involved in establishing cause–effect relationships**. Complications include the long latency periods typically involved between actual exposure to a substance and the production of a tumour (a process, known as carcinogenesis, which may take as long as forty years in some cases). Connected with this is the often substantial problem of pinpointing specific causative agents when the individual may well have been exposed to a whole array of carcinogens. This problem is exacerbated by phenomena such as co-carcinogenesis or synergism whereby two or more chemical agents may act together so as to produce a cancer-causing effect. Difficulties of this type must be considered in the context of the relative ignorance which currently exists regarding the biological mechanisms of cancer development. Taken together, these factors represent considerable difficulties for the initial identification of carcinogenic substances and, perhaps equally importantly, for the subsequent marshalling of sufficient unambiguous evidence so as to convince possibly sceptical observers, including those who have a financial stake in the continuing production of a suspected chemical substance.

It is also important to realize that the technical expertise which relates to chemical carcinogenesis is divided into three main branches. **Epidemiology** represents an attempt to accumulate actual human evidence of excess cancer rates (taken from statistics of mortality and morbidity) and to link these to specific causes. Particular problems for epidemiological studies include the necessarily long period before human evidence can be reasonably collected. Quite obviously, epidemiological evidence can only be amassed *after* undesirable effects have occurred, and even then it may be impossible to relate particular effects back to a specified carcinogen.

Animal Testing avoids some of these difficulties by concentrating on creatures with a considerably shorter lifespan than humans and by actively administering dosages of a suspected substance so that the complexities of identifying causative agents are substantially reduced. On moral grounds, this form of laboratory testing is limited to non-human victims, although the extent to which similar moral considerations should also be applied to animals is open to debate. The problem which remains is that of extrapolating given effects in animals under laboratory conditions to human beings.

Short-term tests provide a third source of evidence with regard to carcinogenicity. The most widely used of these is

known as the 'Ames test' (after its originator, Dr Bruce Ames
of the University of California, at Berkeley). Strictly speaking,
the Ames test is not a test of carcinogenic properties but
rather of the capacity of a chemical or other substance to
cause genetic mutations in a strain of bacteria. However,
a large amount of evidence exists to suggest a strong link
between such mutations and cancer induction. The advantage
of techniques of this type is that they can be performed in
the laboratory at comparatively low cost; the major disadvan-
tage is once again the inherent difficulty in relating such
evidence to human cancers.

Already it can be seen from this brief discussion that there
are immense technical difficulties involved in producing
absolute proof of a substance's carcinogenicity. There is an
unavoidable element of uncertainty, although this uncertainty
will vary enormously between that surrounding a new substance, the knowledge of whose characteristics must be less
than complete, and that relating to a substance such as asbestos whose carcinogenic properties have been recognized since
the early part of this century. However, as can be seen in the
continuing controversy over cigarette smoking and lung
cancer, even the accumulation of enormous amounts of
evidence can be insufficient to convince all interested parties.
In circumstances where absolute proof is rarely established,
the issue becomes one of establishing *sufficient proof*, that
is, evidence which is sufficiently clear to allow a fundamentally conservative approach towards protecting the health of
citizens to be adopted. Unfortunately, what might constitute
'sufficient proof' in a given case is hardly less problematic
than the determination of 'absolute proof'. Later in this
chapter we will examine how this problem is dealt with in
practice.

Another major aspect of technical expertise which should
be emphasized here is the debate over **environmental causation**.
Whilst it is generally accepted that the majority of cancers are
due to agents 'external' to the victim, there is a scientific
controversy over the proportion of cancers which can be

linked to specific 'environmental' factors such as tobacco, alcohol, sunlight, diet, drugs or occupational exposure.

It has become clear as a result of this controversy that the consensus of opinion among technical experts in Britain is rather different from that in the United States. In the USA, there is a general acceptance that 20—40 per cent of cancers may be linked to occupational exposure to chemicals, with 70—90 per cent of all cancers being due to environmental agents of one form or another. Scientists such as Professor Samuel Epstein of the University of Illinois in Chicago have interpreted the apparently overwhelming extent of environmental causation as conclusive evidence that cancer can be almost entirely prevented by the stricter control of suspected chemicals.

In Britain, regulatory authorities have accepted much lower estimates of the extent of chemical carcinogenesis — perhaps 1—5 per cent of all cancers being linked to occupational exposure, with a similarly small proportion for other forms of chemical causation. However, British scientists such as Richard Peto of the Radcliffe Infirmary, Oxford, have placed far greater emphasis on other factors, notably tobacco smoking and diet. At the present time this debate between those who see cancer as largely attributable to involuntary exposure to carcinogenic substances and those who see it as mainly related to the 'chosen lifestyle' of victims shows no sign of abating. Here we have an excellent example of national policies having to be established despite the existence of strong disagreements between technical experts. In the 'Further reading' section of this chapter, reference is made to the interesting exchange on this point between Epstein and Peto in the scientific journal *Nature*.

The controversy over carcinogenesis as discussed here has illustrated the uncertainties which presently exist among technical experts regarding the identification and assessment of cancer causing chemicals. In the next section, we will consider how national policy-making has actually come to terms with these difficulties and also how technical evidence has been balanced against economic and political arguments in reaching regulatory decisions.

Government procedures for the control of carcinogenic chemicals

One of the most striking features of the British regulatory structures as they currently exist is the sheer complexity of the decision-making processes which have been established. Responsibility for making recommendations on carcinogen safety is divided between a whole range of advisory committees corresponding to separate contexts of exposure. The key advisory bodies, their areas of operation and the government departments which oversee their activities are all summarized in Table 1.

Whilst the terms of reference and the structure of these bodies vary considerably, the majority of the groups in Table 1 share certain tasks in common. These can be divided for purposes of explanation into three categories: 'Problem formulation'; 'Decision-making'; and 'Policy execution'.

If we examine the stage of **Problem formulation**, it is evident that there are large differences in the way that each advisory body determines whether a particular chemical entity merits closer examination. It will be appreciated that no action can follow unless a problem has emerged and been placed on the body's agenda.

Apart from the Hunter Committee, whose terms of reference concentrate solely on the health hazards of tobacco, a broad distinction can be made between those which systematically review every available chemical and those which deal selectively with particular substances in a more *ad hoc* manner. For example, the Committee on Safety of Medicines (CSM) considers *every* pharmaceutical product before its release onto the market. The Department of Trade, however, has no such procedure and will only deliberate over a particular consumer produce once questions have been raised about its safety.

Turning now to the **Decision-making** category, four basic

Table 1

Exposure Context	Advisory Body	Government department
Workplace	Advisory Committee on Toxic Substances (ACTS)	Health & Safety Commission Department of Employment
Food	Food Additives & Contaminants Committee	Ministry of Agriculture, Fisheries and Food (MAFF)
Medicines	Committee on Safety of Medicines (CSM)	Department of Health & Social Security (DHSS)
General Environment	Department of Environment, etc.	
Pesticides	Advisory Committee on Pesticides (ACP)	MAFF
Tobacco	Hunter Committee	DHSS
Consumer products	Department of Trade	
General	Committee on Carcinogenicity (COC)	DHSS

issues need to be considered. First of all, there is the question of *membership*. With only one exception, the bodies consist of technical experts chosen on the grounds of their special knowledge of particular problem areas. The exception is the Advisory Committee on Toxic Substances (ACTS) which has special responsibility for the control of workplace carcinogens.

In keeping with the practice of its superior body, the Health and Safety Commission, the membership of ACTS is selected on a strictly tripartite basis. 'Tripartite' in this context refers to the representation of three parties: industry, trade unions and government (both local and national). Accordingly, ACTS does not constitute an expert forum but one in which an explicit bargaining process is carried out between the chosen representatives of different social groups. Of course, technical advice is available to each of the groups, but most have chosen to select proven industrial negotiators rather than scientific researchers as members.

The contrast between the 'expert' and 'representative' forms of advisory committee is especially interesting because of the difference in regulatory philosophies which it suggests. The former model implies that decisions regarding physical risk (whether carcinogenic or otherwise) are 'best left to experts'. Arguments for this model would include the claim that only the best available technical advice can ensure that all the relevant evidence on a hazard is properly considered. A corollary to this would be the suggestion that non-expert bodies have a tendency to over-react to emotionalism or political pressures and therefore become, in some sense, unbalanced or irrational.

The counterclaims in favour of the 'representative' model can only be summarized briefly here. Firstly, it can be argued that the potential victims of any carcinogenic risk should have the right to participate in such processes purely on democratic grounds. This argument, of course, raises questions about both the flexibility and the desirability of public involvement in technical decision-making. This will be considered further below.

The second major argument is that decisions over risk are not simply a question of assessing the physical properties of a potential hazard, they also involve a series of judgements regarding the *social acceptability* of that hazard. In a society which acknowledges that certain physical hazards must be tolerated and that 'zero risk' cannot be achieved, the question becomes one of balancing risks against their likely benefits (perhaps in terms of economic advantage, health improvement, convenience, pleasure value, etc.). The case for the 'representative' approach is that narrowly qualified technical experts are not the appropriate group to carry out a 'balancing act' of this sort.

One other general point which relates to the discussion of risk acceptability is the question of whether or not the various advisory bodies take questions of *need* explicitly into account, i.e., do they consider only the safety of a substance or do they also deal with matters of desirability, necessity or utility? On this issue again there is little uniformity between the groups.

Although most of the bodies do consider 'need' in some way, the definition of this varies enormously. For food additives, for example, there is a 'permitted list' and a manufacturer must produce good reasons why a new substance should be added to it (criteria would include 'consumer demand', 'better performance', 'technical improvement', etc.). Some committees attempt to avoid questions of 'economic need' such as the likely effect on employment of banning a substance, or possibly consider this issue only indirectly. The point is, however, that, since many of the committees do not publish the reasoning which led them to their final decision, it is often extremely difficult to ascertain just how questions of need were considered (if at all). Here we come up against an important feature of the British system: it is essentially a *closed* decision-making system.

A third point of interest within our Decision-making category is the fact that most advisory committees have not been established to deal with *carcinogenicity as a special problem*. In Britain, carcinogenicity is officially looked upon as only a part of the more general assessment of the risk of toxic chemicals. This is in contrast to developments within the United States towards 'generic' cancer policies in which separate criteria are laid down for assessing carcinogenicity. The rationale for this is in terms of the special technical difficulties involved in the identification of a carcinogen (see

above). The US approach is to adopt uniform testing procedures rather than to have, as in Britain, a 'case by case'
strategy for evaluating suspected substances.

Finally, within Decision-making, we must mention the
question of whether or not the functions of *sponsor* (of a
particular industry) and *regulator* (of the same industry)
are combined within one government department: there may
on occasion be a conflict of interest between these two roles.
A notable example here is the Ministry of Agriculture, Fish-
eries and Food (MAFF) which is responsible for the promotion of food manufacturing and agriculture and yet must
also monitor the safety of the relevant products as well as that
of the pesticides which are used to achieve greater productivity.

The third of our broad categories is **Policy execution**, i.e.,
the way in which decisions are actually put into practice once
they have been recommended by the advisory body and
approved by the appropriate Secretary of State. Once again,
a wide variety of practices is found. Broad differences can be
seen in the form of the *regulatory output* produced by the
bodies (recommendations, guidelines, licences, etc.). In the
context of occupational safety, Threshold Limit Values
(TLVs) are of particular significance. A TLV is that concentration of a substance below which it is thought that nearly
all workers can be repeatedly exposed without adverse effect.
The key word here is, of course, 'nearly', as it is still possible
for a small proportion of workers to suffer from occupational
illness despite adherence to the TLV. The *enforcement
measures* which are utilized are especially important because,
without these, regulatory policies clearly have little value.
Differences can also be seen between the bodies in terms of
the *monitoring* of the effectiveness of decisions after their
implementation.

So far in this section our discussion of the British control of
carcinogens has stressed the points of difference between the
regulatory authorities and the wide variety of assumptions and
criteria according to which they operate. However, a number
of factors which serve to provide a degree of co-ordination and

consistency should not be ignored. One of the most important of these is the network of civil servants who form the secretariat of the advisory bodies but who are also in a position to communicate with committee members and officials in other bodies so as to ensure that no wide discrepancies exist in the overall policy output of different groups. The particular function of the Committee on Carcinogenicity (COC) should also be noted. As the central source of technical expertise on matters of carcinogenicity, the COC is in a position to monitor all regulatory activities in this broad area. Through its formal and informal connections with other relevant committees, it assists in the maintenance of consistent decision-making. One other source of consistency may also stem from the technical experts who constitute most of the advisory bodies − British carcinogenicity researchers are a relatively small group and professional links outside the advisory bodies may be a unifying factor. The full complexity of the political processes involved in controlling carcinogens cannot be conveyed without reference to the main non-governmental organizations which are also involved in policy making. These can be divided into three types: industry-based associations, trade unions, and 'public interest' pressure groups.

Sectors of industry have tended to band together for general lobbying purposes so that organizations like the Chemical Industries Association (CIA) or the Association of the British Pharmaceutical Industry (ABPI) have presented a united front whenever possible. In general terms, the CIA and similar bodies have argued for a cautious approach to new regulatory actions in which the costs of any new control measure are taken very seriously into account and no regulatory moves are made without a solid empirical basis. Given this overall concern for the economic viability of industry, it is hardly surprising that organizations like the CIA keep a close watch on regulatory developments and are eager to marshal technical resources of their own (generally through member companies) in opposition to what they consider to be untimely or inappropriate action.

In this respect, one major area of debate in Britain has been over the effect of future and existing European (EEC) legislation on the British control of hazardous chemicals. For example, a notification scheme for new chemicals will eventually be introduced as a response to EEC activity. This scheme will include requirements for the assessment of carcinogenic properties and will be relevant to the protection of both the safety of the workplace and the natural environment. Other important EEC activities relate to occupational health and safety, the general environment, pesticides, pharmaceuticals, food additives, toys, cosmetics and product liability. During the 1980s, European developments of this type will have a significant impact on British procedures for carcinogen control.

One other group which has been engaged in discussions over British and European regulation is the trade union movement via the Trades Union Congress. In addition, whilst single issues like the continuing controversy over the pesticide 2, 4, 5–T or the use of benzidine-based dyes have attracted particular media attention, a number of unions have developed general policies for the control of carcinogenic substances in the workplace. Three unions are especially noteworthy in this respect — the Association of Scientific, Technical and Managerial Staffs (ASTMS), the General Municipal Boilermakers and Allied Trades Union (GMBATU) and the National Graphical Association (NGA). Each of these has reacted to the United States debate over the extent of occupational causation and the need for generic cancer policies by developing comprehensive demands for the conduct of British regulatory approaches to carcinogenic substances. The ASTMS document — *The Prevention of Occupational Cancer* — provides a useful summary of these arguments. Although the particular points put forward by these unions cannot be discussed here, the overall case is that occupational carcinogens have not been adequately controlled in Britain and that firmer action is necessary both at the level of the individual workplace and nationally. Underlying these demands is the belief that this is essentially a *political* issue in which workers have a right to demand that unnecessary cancer risks should not be imposed upon them.

The third category of non-governmental organizations consists of the pressure groups which have sprung up to lobby on issues of carcinogen control. However, in contrast with the United States, these groups are few in number and have remained on the periphery of the policy-making process. In short, chemical carcinogenesis has not been singled out as a major political issue except by newly established organizations such as the Cancer Prevention Society.

Discussion

The starting point for this chapter was the question of how societies can best make sound decisions regarding the control of technological risks. We have used chemical carcinogenesis as a specific focus of attention within this wider problem area. In this last section, I discuss just one important aspect of this general question: the distinction which has been made between 'expert' and 'representative' forms of advisory body. To express the underlying point in somewhat more explicit terms: how might the encouragement of broader public participation in decisions of the sort described above serve to strengthen or, alternatively, weaken the present regulatory system?

We have already referred to some of the arguments on this question. The 'expert' model is seen to have a greater potential for 'rationality' and 'objectivity' and also for the thorough consideration of the available technical evidence. The 'representative' model can claim to be more democratic and also to allow a more effective sifting of the costs and benefits involved in any regulatory action. How might these differences between the models be resolved in practice? One response to this difficult question might be as follows.

Perhaps the first step is to recognize that, as we have seen with carcinogenic chemicals, the present system is heavily weighted towards the side of *expert* committees operating on a relatively *confidential* basis. In this way the general public is excluded not only from the membership of advisory bodies but also from the possibility of scrutinizing their discussions. However, we have noted that these committees cannot avoid reaching conclusions which are not only 'technical' in character but also involve an array of economic, political and social considerations. These considerations include such factors as the cost of regulation, the probable impact on employment, the need which currently exists for a substance and the public perception of a chemical as a significant hazard.

Given the nature of the decisions which are being taken, one can legitimately ask whether a solely expert forum is the best place for these evaluations to be conducted. In what way is a narrowly-trained scientist an appropriate judge of the types of consideration listed above? In addition, might there not be value in dispelling much of the secrecy from the system so that advisory bodies need to explain the basis on which decisions have been reached? A change of this sort would permit outside groups at least to challenge policy recommendations by replying directly to the key assumptions and arguments involved. In the United States, public hearings are held on all decisions which are likely to prove controversial. Oppositional groups thus have the opportunity to make their view generally known and decision makers are forced to justify their conclusions to a wider audience.

Our discussion of public participation has so far identified two important aspects of this issue. Not only might it be in the public interest to broaden the membership of advisory bodies but also to allow greater external scrutiny of decision-making procedures. It might be argued in opposition to such a viewpoint that members of the general public are unlikely to hold worthwhile views on, in this case, specific chemical structures and that such representation may even prove a hindrance to calm, careful discussion. A number of points may be made in reply to this argument.

Firstly, while it can be accepted that lay members may not be in a position to comment extensively on the technical minutiae of every suspected hazard, they may be qualified to offer an opinion on extra-scientific considerations such as those discussed above. Secondly, it may on the other hand be the case that the representatives of groups who are likely to be affected by any present or future hazard might well study the scientific detail in order to protect their own interests. For this reason, non-expert decision-makers would not necessarily be totally uninformed on technical matters. Thirdly, a case could be made that public participation can most usefully be employed at the level of drawing up general regulatory guidelines and requirements for the balancing of costs and benefits. In this way, a general framework could be established by a 'representative' body (which might include a number of technical experts) but the specific implementation of this would be left to expert committees which would be finally accountable to the policy-making body. For example, a 'generic carcinogens policy' might be drawn up via participatory procedures, with the advisory committees discussed in this chapter having the authority to make specific recommendations within this. However, this is just one suggestion. If the principle of representation is accepted then there is a need for further consideration of how this principle might be transformed into practice in specific areas of decision-making.

In this chapter, I have attempted to stress the importance of the *procedures* according to which decisions over technological risk are taken. If, as certainly seems to be the case, there is no such thing as an *intrinsically* acceptable level of risk then the best we can hope to achieve is a level of risk that has been decided upon by the most equitable methods available. Hence the significance of such issues as the parts to be played by technical expertise and public participation. It may well be that the evaluation of the physical costs of technical change is too important a matter to be left to experts or indeed to any single group of policy makers. If that is true, then urgent thought needs to be given to how the franchise for technical decisions is to be extended.

Questions for discussion

1. Is it possible for the general public to play a meaningful part in technical decision-making?

2. How might the British system for the control of carcinogenic chemicals be significantly improved?

3. Should decisions of the sort discussed in this chapter be taken in full public view or behind a veil of secrecy?

4. What part should technical experts play in the assessment of physical risk? What problems are there for experts in assuming the role of government advisers?

Further reading

L. Doyal *et al.*, *Cancer in Britain* (Pluto Press, London, 1983).

This publication reviews the debates over chemical carcinogenesis and examines a number of case-studies of carcinogens in the workplace, consumer products and the general environment. It then advocates a broad strategy for reducing cancer deaths in Britain and elsewhere. The book contains several useful appendices − including a full reprint of the exchange between Epstein and Peto in *Nature*. For a contrasting analysis of cancer prevention, see;

R. Doll and R. Peto, *The Causes of Cancer* (Oxford University Press, New York, 1981).

This book has been extremely influential both in the United States and Britain. It is written in a relatively accessible fashion and provides an introduction to the technical literature in this field.

ASTMS, *The Prevention of Occupational Cancer* (1980). Available from ASTMS, 10−26a Jamestown Road, London NW1.

This publication provides a clear account of one trade union's case for the stricter control of chemical carcinogens. Following extensive criticism from manufacturers' associations. ASTMS produced a restatement of their case:

ASTMS, 'Prevention of occupational cancer — again' *Health and Safety Information, Monitor No. 11* July 1981. Available from ASTMS.

S. Gusman, K. Von Moltke, F. Irwin, and C. Whitehead, *'Public Policy for Chemicals — National and International Issues'* (The Conservation Foundation, Washington, D.C., 1980).

This book provides a reasonably thorough account of the notification, testing and control of toxic chemicals with specific reference to Western Europe and the USA. It is recommended for those wishing to understand the finer detail of contemporary regulatory processes.

6

IS SCIENCE INDUSTRIALLY RELEVANT?
THE INTERACTION BETWEEN
SCIENCE AND TECHNOLOGY

M. Gibbons

Introduction

One of the aims of the study of Science, Technology and Society is to understand the nature of scientific activity in all its historical complexity. Another aim is to use the knowledge so gained to improve the linkage of scientific activity to socially useful purposes. In other words, STS, as with science itself, has both a theoretical and a practical aspect. It is, however, important to remember that neither science nor the societies in which it flourishes are static entities and, therefore, our understanding of scientific activity has a historical dimension.

One of the key insights derived from sociology is that if a group is to survive in society it must receive support from, or be legitimated by, the other social groups that are dominant in that society. This was the situation of science at the beginning of the scientific revolution. Science — the study of nature, mathematically — needed the support of princes, academia and the church. It has been argued that Galileo realized very clearly that if his activity was to survive and grow he would have to enlist the support of the Church. While one might disagree with how he went about this task or debate whether he was the ideal person to carry it out, the problem is recurrent. Throughout its history, science has had to explain its purpose to the powers that be and, because the interests and concerns of those powers have changed historically, so science has had to find different ways to express what it was doing if it wished to survive.

From the outset, one constant element in the armoury of those who spoke for science has been its utility. But it is important to remember that at the beginning, science had very little in the way of practical achievements to offer. Through most of the sixteenth and seventeenth centuries, statements about the utility of science were essentially statements about future intentions, rather than either past or present, achievements. Nevertheless, the association of Galileo with the engineers of the Venetian arsenals and the involvement of Newton with the problems of mining and navigation have been used to great effect to ensure that the utilitarian aspect of science was not forgotten. However, it was to take some time before science was able to deliver the goods, but as the First Industrial Revolution (1750–1850) got underway, it looked very much as if things were going to change. The spinning jenny, the steam engine, the manufacture of Wedgwood china and many similar achievements came to be interpreted as the results of the application of science. Until very recently, there was hardly a serious history of the First Industrial Revolution which did not pay at least residual respect to the founding fathers of modern science for providing the methods, techniques and theories upon which the vast network of technical apparatus apparently depended. It is perhaps worth pointing out, however, that since the end of World War Two a vast amount of scholarship has been devoted to the question of the role of science in the First Industrial Revolution. Inevitably, schools have emerged taking one side or the other in the debate, and, as so often happens, scholars have been forced to clarify what they mean by science; to distinguish scientific discoveries from the application of scientific method; to separate empirically verified associations between scientists and industrialists from the larger claim that the discoveries of science were actually used in industry. For example, so great an authority as David Landes has concluded that the inventions on which the First Industrial Revolution depended owed more to the work of 'inspired tinkerers' than to the discoveries of science. Expressed in contemporary terms, the First Industrial Revolution was more the result of technological than scientific inventiveness, although it must be recognized that science then was a smaller, less differentiated, enterprise than it is today.

If the debate about the part played by science in the First Industrial Revolution is marked by sharp disagreements, judgements about its role in the Second Industrial Revolution (1850–1914) are marked by consensus. According to most historians, the social history of science in the period after 1850 is characterized by the emergence of a disciplinary structure not very different from the one we know today; for example, it had become possible to distinguish, within what had previously been called natural philosophy, autonomous centres of research activity labelled physics, chemistry, the geological sciences, etc. In addition, and proceeding along with this disciplinary differentiation, it was becoming possible for people to make a living by practising science. During the period 1850–1914, first in Britain and western Europe and then in the United States, science was becoming a fully professional activity and, so the argument runs, the relationship between science and technology, between scientists and technologists, grew more intimate. The great industries of the Second Industrial Revolution, particularly those based on chemicals and electrical machinery were the more or less direct result of prior discoveries in chemistry and physics and would not have been possible without them.

Such conclusions seem plausible enough because they conform to what is virtually an unquestioned general belief that ideas are the currency of progress. It is but a small step from such a belief to the view that the theoretical discoveries of science form the basis of technological and, hence, industrial progress as well. So deeply held is this view about the relationship between ideas and progress, that it comes as something of a shock to have to consider whether things might be otherwise.

That the situation may be more complex than had previously been thought may be illustrated by the case of thermodynamics. The subject is in many ways an ideal of scientific exposition. Thermodynamics is an axiomatic, deductive system and virtually all the energy processes of both closed and open systems can be derived mathematically from a few simple principles. Yet, the pioneering theoretical work of Carnot on heat engines was produced many years after the technologists had launched, and entrepreneurs built, the steam engines that Carnot must have seen operating in France. In fact, it appears that it was the existence of working steam engines that led Carnot to develop a theory about them and to try to determine what increases in efficiency might be attained. To express this situation in another way, it could be said that, in this case, *technology presented a problem for science*. One may then be prompted to enquire how often this has been the case and to question more generally one's views about the relationship between theory and practice. It might be worthwhile to point out here that the 'liberal view' of the relationship of ideas to progress is currently being challenged by a 'materialist' conception which is derived from the writings of Karl Marx and Friedrich Engels. The challenge is serious because the views of the one are diametrically opposed to the other.

Science and technology in the twentieth century

This brief sketch of the relationships of ideas to progress and of theory to practice may be sufficient to show that the question of the links between science and technology is a historical one. What precisely these links are depends critically not only on whatever view of progress one may hold, but also on the historical period in question. That is, the answer to the question will depend on the level of development of industry, of technological skills and practices and, lastly of science itself. As we have already seen, the view of many historians is that scientific and technological activities are moving into a closer, perhaps even a symbiotic, relationship. But what does this mean?

In the twentieth century, as opposed to the eighteenth or the nineteenth science has become a thoroughly professional and specialist activity. Not only can men and women now make a living through the practice of science, but they can do so in any one of a number of different disciplines. Technology, too, is differentiated both intellectually and socially from science and, although technology may be said to have appropriated the methods of science, it none the less claims for itself a certain autonomy of subject matter. Technological activities are also, of course, both professionalized and specialist in their own right. It should be noted further that scientific and technological activities are currently carried out in different social institutions. Science aspires, where it can, to the freedom of universities while technology usually evolves within the social framework of industry in general and the business enterprise in particular. But, of course, there are scientists in universities who have contributed to technological projects and, conversely, technologists who make contributions to the scientific literature. The upshot of all this is that the traditional language which distinguishes science clearly from technology is beginning to break down. When we attempt to distinguish between pure and applied science, between short-term versus long-term research and between curiosity-oriented and mission-oriented research and development we merely bring to light the cognitive, institutional and psychological *dimensions* of two closely related activities.

It is perhaps worth illustrating some of the difficulties that one gets into if one tries to get a 'clear and distinct' idea of what science and technology are, and of the differences between them. For example, how would one classify physicists, working in a university, on a problem relating to the structure of matter? Clearly, it could be said that they are doing pure science. But, if we were told that the specific problem was concerned with the resistivity of silicon and that it was hoped that the results would be of assistance in the design of a solar antenna for a satellite, should not this work be properly termed applied science? On this reading, the distinction rests on whether or not there is an application in view. Now, would the situation be modified if the scientists themselves *said* that in undertaking the work they had no application in mind but, in any case, it would be a long time, if ever, before such an application became likely? In other words, do we get any help from knowing that scientists are working on long-term as opposed to short-term problems; and would their work need to be reclassified if all of a sudden an application emerged? Some have sought to escape this dilemma by defining research according to the *motives* for which it is undertaken. If scientists say that they are mainly interested in the phenomenon of resistivity, they could be said to be doing curiosity-oriented research; if they are consciously pursuing research in the direction of a goal (e.g., a solar antenna) they are doing mission-oriented research. A further clarification might be that universities are the places where curiosity-oriented research is supposed to be carried out and industry and government laboratories the places where mission-oriented research is supposed to be carried out. This suggests, perhaps, that science might also be classified according to the institution in which the research is located. But how would one classify scientists who are working in a university laboratory on a problem relating to the structure of matter, the long-term aim of which is to contribute to our understanding of solar antennae but which is funded by the Ministry of Defence? Presumably, according to our definitions, the scientists are working on pure, long-term, mission-oriented research – or is it applied long-term curiosity-oriented research? One can permute these definitions of science among its various cognitive, institutional and psychological dimensions and, although it is possible to get silly combinations, no one set of definitions is very satisfactory. Indeed, the ideal of completely disinterested research, which as we shall see below is of contemporary policy interest, is itself a product of the way science used to be (or at least how we thought it to be) in the eighteenth and nineteenth centuries and, alas, may be disappearing.

So far, attention has been focused on science but, equally, technology is difficult to define. Indeed, it is possible to ask whether research carried out in university engineering departments is pure or applied, is short- or long-term, and whether it differs in any significant way from research carried out in the laboratories of government and industry. In the end, one has a not very satisfactory set of definitions of highly idealized activities which often reveal more about the views of the people trying to define them than the activities they are trying to describe.

But, you may ask, why all the fuss about describing and defining scientific and technological activities? The short answer is that, despite enormous successes during this century, science is once again under scrutiny. The scientific enterprise of virtually every developed nation has, in the last twenty years, been brought before the bar of public accountability and has been required to justify the large amounts of public funds which are now spent on scientific activities. One of the strategies of the scientific community has been to try to establish that a close link exists, via technology, between science and its application to economic ends. It is a strategy that had been used very effectively throughout the nineteenth century in Britain and the United States, but which was not needed during the twentieth, possibly because of the successful performance of scientists in the two world wars. From 1960 onwards, however, the slowing of economic activity in the West has been coupled with a much closer scrutiny of public expenditure generally. It was in this context that the question of the economic relevance of science – particularly in its probably now non-existent form of pure, long-term curiosity-oriented research – was once again re-opened.

The strategy adopted by scientists was to demonstrate that apparently curiosity-oriented work would, in the fullness of time, lead to important money-making technological innovations. Although there is, in the pronouncements of the leaders of the scientific community, no shortage of testimony to the belief that the dispassionate research of earlier generations has yielded useful results, the empirical evidence necessary to drive the point home has seldom been presented. Contemporary rhetoric is strewn with examples which make the point as if it were demonstrated. Examples include penicillin, radio and television, X-ray machines, various types of antibiotics, lasers, etc. But rarely is the connection between the final product and the underlying scientific contributions made explicit. But, you may ask, how could such links be made explicit? How does one establish empirically the contribution made by scientific knowledge to new products? These are far from being trivial questions and the search for answers is complicated by the fact that science itself has been developing throughout the last two centuries and, therefore, we may expect that the ways in which it has contributed to technological development have been changing as well. In addition, one would need a good reason to undertake such a study. After all, there is the possibility of a negative or equivocal result which, in the hands of those hostile to science, might be used to undermine public support for science.

During the late 1960s and throughout the 1970s, science came under attack in varying degrees in the countries of the Western world. While most of the public reaction was concerned with the uses to which science was being put and to the involvement of universities in the military–industrial complex, the political reaction was directed at the escalating costs of science and at the economic benefits derived from public investment in science. During this time the generally accepted belief that science provided the basis for future generations of technology came under close scrutiny and, to the surprise of many, precise empirically verifiable links between science and technology were proving difficult to find. It was in this context that the National Science Foundation in the USA commissioned a study with the aim of providing clear and unequivocal examples of the ways in which science has contributed to the development of new technology. It is worth spending a little time on this study, not only because it is representative of a genre but also because it illustrates clearly the difficulty of providing evidence on this apparently simple matter. In a report entitled *Technology in Retrospect and Critical Events in Science* (TRACES), a team from the University of Illinois set out its analysis of five innovations (that is, examples of new products) which have had a clear and unambiguous economic importance. One of these is the development of the oral contraceptive pill and we shall examine this case study here because it illustrates clearly many of the issues which we have touched on so far.

First of all, as with many other examples of empirical research, the selection of data must be guided by a model of some kind. In this case, the investigators have assumed that the history of a new product can be considered to be composed of a number of discrete events. These events are subsequently classified into three types: curiosity-oriented research, mission-oriented research and development events. It is further assumed that these events can be clustered into streams broadly related to the existing disciplinary structure of science. In the case of the contraceptive pill, for example, the three disciplinary streams identified were hormone research, the physiology of reproduction and steroid chemistry. The disciplinary structure of science thus serves as a skeleton on which to hang various types of events. An event is defined as the discovery, whether theoretical, experimental or technical which, in retrospect, is perceived as essential to arrive at the final event − the contraceptive pill. The results of the analysis are presented in Figure 2.

From the figure one can grasp at a glance the prehistory of the Pill reconstructed as a series of events flowing through time down the disciplinary streams and culminating in the final product. All one has to do now is to count up the number of events:

Type of event	No.	
Curiosity-oriented research	33	(63.5%)
Mission-oriented research	15	(28.8%)
Development events	4	(7.7%)

Figure 2 The development of the oral contraceptive (*overleaf*)

A simplified 'family tree' showing the oral female contraceptive to be the culmination of decades of research, much of it non-mission-oriented, in initially unconnected areas.

Two important sets of events pre-dated the work of the 1920s to 1960s: male and female hormones were discovered and the functions of the corpus luteum described between 1849 and 1921 (**5**); and steroid chemistry was initiated (1903 (**18**)).

Reproductive physiology developed from 1922 (**6**) when Evans and Long speculated on the existence of a hormonal feedback mechanism controlling ovulation. Progesterone was identified as an active component of the corpus luteum in 1929 (**7**) and, following the discovery of the potency of crude hormones (**16**) progesterone was isolated and extracted from the corpus luteum (**17**).

In steroid chemistry, the chemical configuration of cholesterol was determined in 1929 (**19**) and female hormones were isolated, crystallised, refined and concentrated (**20, 21**). As a result of **20, 21** and **16** the first mission-oriented research and application took place at Schering AG and Parke-Davis Inc. (**22, 23**) leading to the manufacture of hormones from animal sources. By 1934 four distinct groups in Switzerland, Germany and the USA had reported the isolation of pure progesterone from animal sources (**24**).

Medical research during the 1930s clarified the effects of oestrogen and progesterone on ovulation and the menstrual cycle in animals (**8**) and subsequently brought the first investigations of hormones to treat disorders of the human menstrual cycle (**9**). Papers by Taylor in 1931 (**1**) and Kurzrok in 1937 (**2**) proposing the use of hormones in temporary male sterilisation and contraception influenced this and subsequent work greatly.

In the field of steroid chemistry, meanwhile, the relationship between cholesterol and female hormones was demonstrated (1931), cholesterol was prepared from plant sources (1935) and finally cholesterol was converted to a large number of sex hormones (1936–41, **25**). These achievements led to the formation of Syntex Inc. to produce sex steroids in 1944 (**26**). By this time, Albright and others (**3, 10**) had shown that oral preparations of oestrogen and progesterone might be used to treat menstrual disorders and control fertility. As a result of a crucial meeting between Sangster, Pincus and Charig in 1951 (**4**), full-scale evaluation of synthetic steroids (**11**) and progesterone (**12**) was begun and this work influenced Syntex Inc. and Searle to synthesise progestin (**27, 28**).

In 1956 mixed-hormone preparations from Searle (**29**) and Syntex (**30**) were released and subsequently approved for treatment of menstrual disorders (**31**).

Finally, further field studies of synthetic progestins in animal and human populations in 1956 and 1957 (**13, 14**) led to the vital expanded 1959 field studies in Puerto Rico, Port-au-Prince and Los Angeles (**15**) and the full approval of the Food and Drug Administration for the female oral contraceptive.

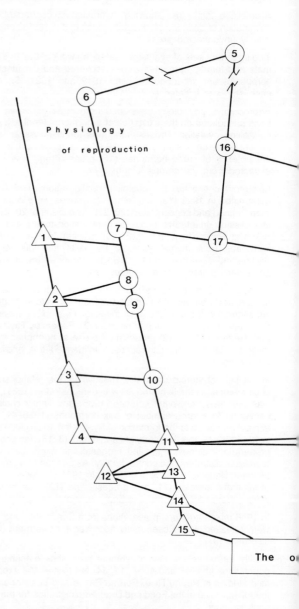

Physiology
of reproduction

The o

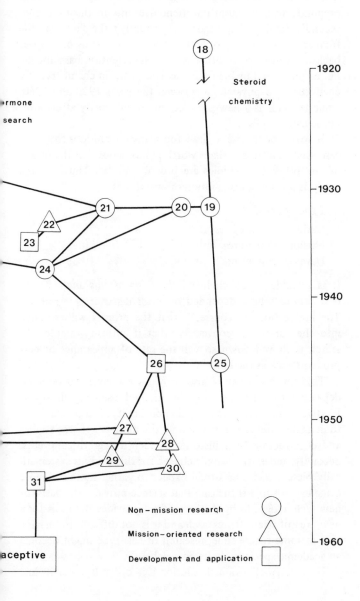

One can note two things from this single example; firstly, the preponderance of curiosity-oriented events in the total and, secondly (and perhaps more importantly) the fact that the frequency of curiosity-oriented events increases as one moves backwards in time. In other words, the scientific basis for the Pill as illustrated in Figure 2 was laid down in chemistry, bio-chemistry and physiology between the years 1920 and 1950, while the Pill, as a new product, appeared shortly afterwards, in the early 1960s.

It would be unwise to draw too many conclusions from one case study but it is perhaps worth pointing out that the results of the other case studies are broadly similar. Thus, for the study as a whole a distribution of events was:

Type of event

Curiosity-oriented research	70%
Mission-oriented research	20%
Development events	10%

It is possible to conclude *on the basis of this analysis* that new products have depended on prior discoveries in science. The implication, of course, is that the process will continue into the future or, conversely, that if curiosity-oriented re-search is slowed down so will the rate of appearance of new products such as the Pill.

This single case-study also provides a convenient point of departure for reflection on the method used in this study. Firstly, the method is of general applicability. It is not difficult to imagine multiplying indefinitely the numbers of case studies and constructing their histories by identifying critical events. Secondly, unless the sample chosen for this study is systemati-cally biased, then one would expect to gather more and more evidence to support the case that science provides the basis for new technology. In brief, the method provides the basis for a new area of scientific research and it is not difficult to imagine that at some future time, research of this type might become an academic speciality in its own right!

However, perhaps a little caution is in order. Is it worthwhile extending the methodology to larger numbers of products? Is the method really sound? Although this is not the place for a critical review of methodology, it is nonetheless important to realize that the whole method hinges on being able to identify critical events and to link them together in historical time series. Here, I think, there is cause for concern. While no-one can doubt that the events occurred, the act of linking them in a chain to the Pill requires a retrospective judgement of the relevance of the event, and the sole criterion of relevance is its relation to the Pill. It is possible, for example, to imagine the same set of critical events figuring in another set of linkages to a different product altogether. In other words, by reconstructing the past in this way, one is attributing a degree of specificity to the contribution of a given critical event which requires, to say the least, closer scrutiny. If one contribution were removed, say, Windaus's work at Guttingen in 1929 on the chemical configuration of cholesterol (see Figure 2), would it mean that the Pill would not have been developed?

A second, related point is that no weightings are given for any of the critical events; they are all presumed to be equal. On this reckoning, the contribution of a major scientific discovery is counted equally with what would be recognized as a trivial one. Thirdly, the lines connecting one critical event with another are all simple and direct and, certainly, convey the impression that the one leads to the other. The possibility exists, however, that the two events were historically unrelated — that is that scientist *B* had never heard of, nor was he influenced by, the prior work of scientist *A*. If this were the case what would be the meaning of the line joining the two events? Fourthly, the results of the study are crucially affected by the time period chosen for retrospective analysis. In the cases considered here, a forty-five year horizon was chosen; a different picture would emerge if a ten year horizon were used instead, as can be clearly seen by examining Figure 2. Fifthly, the critical events are presented without context. There is no indication whether any extra scientific factors played a role in any of the events chosen. This is important because the evidence is being interpreted as showing the economic importance of prior scientific research as if the research was motivated by curiosity alone, and that no social, economic or technical factors had entered into it. But, what if the event was initiated by factors unrelated to the Pill, say by a war, an economic depression or social unrest; might not this mean that the scientific event would need to be re-interpreted?

The purpose of the last paragraph is not to denigrate the results of research seriously undertaken, but to drive home a point made in several places in this book (see, for example, Chapter 8) – that the results obtained in the TRACES study depend on *a prior choice of a model* of the relation between science and technology. If one wants to challenge the results it is necessary either to criticize the model or provide an alternative, more realistic one. The difficulty is to think of another way of conceiving the problem because the way of thinking about the relationship between science and technology exemplified by the TRACES study has such a powerful hold over our imaginations. None the less, a few possibilities exist though they may not seem as convincing as the results of the TRACES study which have been summarized here.

Discussion

Consider, for example, the problem from the point of view of the product. If we chose to examine, in detail, not just the development of the Pill in general but the development of a specific brand of pill by a specific pharmaceutical company over a certain interval, what would the relationship between science and technology look like then? Now, the examination of the factors governing the emergence of new products is the proper preserve of the field of activity known as industrial innovation studies, and about the factors which affect innovation a good deal is known. The model which governs much of this research is that new products emerge in the *interaction of a technological opportunity and a market need*. (Note that in the TRACES study no mention is made of market forces.) In the example of the Pill, the specific type of pill, its cost, the timing of its launch onto the market and subsequent sales are best explained in terms of such factors as detailed knowledge of users' needs, investment in research and development (R & D) and, generally, good communications between the various functions — research, production and marketing etc. — of the firms. One possible entry point for scientific discovery is through the R & D function of the firm, but it is important to bear in mind that much industrial research and development is not concerned with what in TRACES was described as curiosity-oriented research, or for that matter, mission-oriented research, but rather with development events. Much, if not most, of industrial R & D is directed to solving problems presented by current production technology or is defensive in the sense that it is aimed at protecting the firm's existing market position. In most firms there is little time or resources available for the general advancement of science.

By shifting our perspective, then, from the scientific sources of a general product development to the factors affecting a specific innovation, the contribution of science is, understandably, diminished. It remains a possibility, of course, that the work of prior generations of scientists now forms part of knowledge generally and, when trying to identify specific factors affecting innovation, this type of knowledge is simply overlooked. For example, it would be hard to argue that the development of the Pill did not draw on the discovery of the periodic table by Mendeleev or that modern machine tools do not make use of Newton's laws of mechanics, but it is doubtful whether studies of innovation would include either the works of Mendeleev or the *Principia* in their discussion of factors affecting innovation in pharmaceuticals or machine tools.

It might be useful to enquire, then, about the origins of technological innovations. Now, the question of origins is, as we have seen in the example of the Pill, a tricky one. It could refer, for example, to the *place* where the idea emerged — that is a university or a government laboratory, or to the disciplinary matrix of which it forms a part — that is, to physics or chemistry or biology. As we have seen, the TRACES study was more interested in the latter than the former. It is interesting to note, then, that empirical research into *the sources of the ideas on which technological innovations are based shows that these lie mainly in previous technology*. In other words, if you examine the technical content of a given product and try to trace its antecedents, rather than finding these in prior science they are more often found in prior technology. This is true, for example, in many mechanical and a host of electrical and electronic products. This might even be true for the Pill but, as we have seen, its technological antecedents could not emerge from the model being used.

From the point of view of our understanding of the relationship between science and technology, these findings are extremely valuable because they suggest an alternative way of conceiving the problem. The conventional way of regarding the relationship between science and technology is, as we have seen, one in which the results of science are transformed into technology; in other words, *technology is applied science*. It is possible, however, that technology is autonomous; that it is a form of knowledge with a life of its own. Thus, it becomes possible to speak of technological discoveries — that is, of new insights into ways of making or doing things which derive their inspiration from the field of possibilities presented by the technology itself. As with all other forms of creativity, technological creativity may often make use of other resources in pursuit of its objectives, and there is little doubt that science is one of these resources. The notion that technology is an autonomous activity also carries with it the implication that it constitutes a body of knowledge. About this there can be little doubt. The mere fact that technological knowledge is not usually systematized and presented as a body of theoretical knowledge is certainly not sufficient grounds for regarding it as of secondary importance to science.

Perhaps enough has been said to indicate the existence of a

model of the interaction between science and technology in which both science and technology constitute autonomous streams of knowledge. These streams flow through time, essentially independently, but because they are not hermetically sealed from one another there are times when they do interact. Thus, for example, there might be times when the developments of a technique in the technological stream facilitate the advance of science, as was the case of X-ray crystallography in the determination of the structure of DNA; or, conversely, when ideas in the scientific stream make possible new technological developments, as was probably the case with the oral contraceptive pill.

A number of important conclusions can be drawn from this new model. Firstly, as we saw in the introduction, historians, in attempting to evaluate the role of science in the First Industrial Revolution, have been forced to conclude that the specific innovations of the Revolution were due to the work of 'inspired tinkerers'. Such findings, though they are often disconcerting to scientists, are covered by the two-stream model presented above. It is not difficult to explain the major innovations of the First Industrial Revolution in terms of the creativity of engineers and designers working with existing technological knowledge. According to the historians, however, the Second and Third Industrial Revolutions would not have been possible without discoveries in chemistry and physics. This may be true, but in no case were the discoveries simply applied. Rather, they were absorbed into the already existing traditions of chemical and electro-mechanical technologies: the discoveries did not of themselves create these technologies. As to whether it is helpful to regard science and technology as drawing closer together as is sometimes suggested, it may be more accurate to regard the two streams of knowledge as interacting more closely – in the sense of exchanging relevant information, insights, theories etc.

But, as we have seen from the very brief discussion of the results of innovation studies, there is no evidence that science is becoming a more important source of ideas on which technological innovations are based. Indeed, in certain areas, notably mechanical and electronic industries, it appears that technological inventiveness is a prime source of new products.

There is a third possibility — a development of the two-stream model — which is currently receiving close attention. If it is true that science builds on prior science and technology builds on prior technology, are there any particular situations in which their propensity to interact is high? Is it possible that during the initial stages of a new development in either science or technology there is a great deal of interaction or cross-fertilization, but that as problems are solved and relevant techniques determined the need for interaction decreases? For example, there is now a great deal of interest in the emerging technology of genetic engineering. This interest arises partly from the existing biotechnology industries, some of which — e.g., fermentation — constitute very old technological traditions indeed. These industries are looking to the science of genetics for ways of improving existing processes, or perhaps even for ways of creating new, more economic biotechnologies. These are exciting times in biotechnology and all the evidence indicates that the flow of geneticists into traditional biotechnology industries has increased enormously over the past decade. The question is, how long will it continue? The suggestion we are making is that some of the insights of genetics may be absorbed into the technological tradition of these industries, which will develop them in the direction they choose. Thereafter, communication with a scientific tradition of genetics may well decline. There is some evidence, for example, in the field of chemical engineering that this pattern has been followed. But there is still much empirical work to be done before one could say, definitively, whether this form of interaction between science and technology is more than a bright idea.

Finally, it should be evident that the choice of model for the interaction between science and technology has implications for public policy. The model of technology as applied science favours a policy of supporting scientific research as a means to increase the flow of new products into the economy, while the 'two-stream model' favours supporting technological and engineering research if the aim is to promote technological innovation. The two-stream model also suggests that an important aspect of public policy should be aimed at keeping the lines of communication between science and technology open. Neither model is, however, sufficiently developed to suggest how or when it would be most appropriate to try to improve this interaction.

Questions for discussion

1. In this chapter some examples were given which highlight the problems involved in determining empirically the nature of the interaction between science and technology and industry. Can you think of other examples which might better illustrate the nature of this interaction? How would you go about identifying a statistically significant set of contemporary innovations for further study?

2. David Landes in his book, *The Unbound Prometheus*, attributes Britain's technological successes during the Industrial Revolution not to scientific discoveries but to the work of 'inspired tinkerers'. In contrast, he argues, the Second Industrial Revolution (that is, after about 1850) clearly depended upon previous developments in the sciences, particularly physics and chemistry. What changes had been taking place in British society which would explain such a fundamental change in the role of science?

3. It is sometimes said that science and technology in the twentieth century have drawn closer together, that they form a symbiotic relationship. What does this mean? Is it something that could be tested empirically?

4. In many countries greater public awareness of the cost of scientific research has made it necessary for the scientific community to demonstrate that economic benefits do derive from scientific research. Do you think that studies like TRACES are useful in demonstrating this? What other kinds of studies make this case better?

Further reading

D. Landes, *The Unbound Prometheus* (Cambridge University Press, 1972).

A clear account of the role of technological change in the industrialization of Western Europe from 1750 to the present.

J. Langrish *et al., Wealth from Knowledge* (Macmillan, 1972).

This study of technological innovation includes many examples of the interaction of science and technology in British industry.

K. Green and C. Morphet, *Research and Development as Economic Activities* (Butterworths, 1977).

Provides an excellent introduction to many of the themes discussed in this chapter.

K. Pavitt and M. Worboys, *Science, Technology and the Modern Industrial State* (Butterworths, 1977).

A useful survey of the changing roles of science and technology in modern society, how they are involved in industrial development and war and what this means for scientists and technologists.

7

TECHNICAL CHANGE

Pier Paolo Saviotti

Introduction

The existence of once-unknown technological artifacts such as, for example, motor cars, computers and aircraft, would by most observers be attributed to technical change. New products and techniques have, of course, been adopted by mankind throughout history, but the rate of this activity increased dramatically in Western Europe during the First Industrial Revolution, which began in about 1750. The widespread adoption of the steam engine, the mechanization of numerous operations which had previously been performed manually, the transformation of the cottage industry and the craft shop into the factory system, were some of the events commonly associated with the Industrial Revolution. The presence of new technological artifacts, the conditions that gave rise to their widespread adoption and the effects that they exerted on human lives form a network, so closely interwoven that cause and effect are, at times, difficult to separate.

What within this network constitutes technical change and what purely organizational change is not always easy to discern. New machines and products, at times based on scientific discoveries, may be introduced as a consequence of economic or social changes. Higher labour or raw material costs, improved banking conditions, some cultural and political changes can induce or provide a favourable environment for the adoption of new techniques and products. To deal with technical change is, therefore, to deal with a phenomenon whose main features stand out clearly but whose borderline with related phenomena is not very well defined. In this chapter, three of the most important questions about technical change will be explored, albeit briefly, and in a way that is intended to be illustrative rather than exhaustive. These questions are: (1) What is technical change? (2) How does it arise? (3) What are its effects?

What is technical change?

A perfect definition of technical change would apply to all the examples of technical change which have arisen during the history of mankind; a more limited approach will be taken here. Two important examples of technical change will be examined in order to draw some general conclusions.

One of the periods of history in which technical change was associated with profound transformations in the way in which people lived and worked was the First Industrial Revolution (1750–1850). The First Industrial Revolution is usually associated with three classes of technological innovations: the mechanization of industries, in particular the textile industry; the adoption of new, inanimate sources of power, the most important of which was the steam engine; and a radical change in the techniques of making materials such as chemicals and steel.

Let us consider some of the developments which occurred in the textile industry. The manufacture of almost any textile may be analysed into four main steps: *preparation*, in which the raw material is sorted, cleaned and combed out so that the fibres lie alongside one another; *spinning*, in which the loose fibres are drawn and twisted to form a yarn; *weaving*, in which some yarn is laid lengthwise (*the warp*) and other yarn (*the weft*) is run across, over and under the longitudinal lines to form a fabric; and *finishing*, which may comprise fulling or sizing, cleaning, shearing, dyeing, printing or bleaching. Let us look more closely at the spinning and weaving stages.

Weaving takes place on a *loom*, a device in which the warp

and weft threads are laid. Alternate warp threads are raised so that the weft can be passed through the gap or 'shed' formed between the warp threads (see Figure 3). When this set of

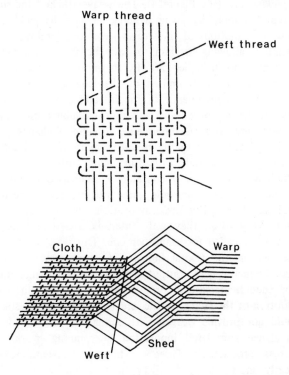

Figure 3

threads is lowered, the other threads are raised and the weft passed back through again. In the looms available at the beginning of the eighteenth century, the weft thread came from a bobbin which was contained in a *shuttle*. The weaver passed the shuttle between her/his hands and through the shed. The rate of weaving was then limited by the rate at which the weaver could throw the shuttle back and forth through the shed, and the width of the cloth was limited by the weaver's arm length. The invention by John Kay in

1733 of the *flying shuttle* revolutionized weaving. The movement of the flying shuttle was not controlled directly by the weaver's hands but through a system of *strings, pickers* and a *picking peg* (see Figure 4). The weaver flicked the shuttle across the loom by giving a quick jerk on the picking peg. The body motions of the weaver were considerably reduced, therefore allowing her/him to pass the shuttle through the shed at a much higher rate. Furthermore, the width of the cloth was no longer limited by the length of the weaver's arms. As a consequence of the introduction of the flying shuttle, the rate of weaving was almost doubled. **Productivity increases are a general consequence of technical change**.

The invention of the flying shuttle by itself would not have caused an industrial revolution. Many other innovations, which accounted for a considerably increased productivity of the economy as a whole, came into use during the Industrial Revolution. But these innovations were not all independent. Very often they were interlinked and at times one innovation called forth another one. This was, for example, the case in the textile industry. Cotton weavers used the thread which was the output of the cotton spinning industry. The sudden increase in productivity in weaving due to the adoption of the flying shuttle, created an imbalance: spinners could not provide weavers with enough cotton thread. This imbalance stimulated innovation in spinning in order to increase productivity. Three very important inventions, which greatly increased productivity in spinning, were made in the following years: James Hargreaves' *Spinning Jenny* (1765); Richard Arkwright's *Waterframe* (1769); and Samuel Crompton's *Spinning Mule* (1771). Some general characteristics of these innovations should be noted. In the spinning wheel, which was in general use at the beginning of the eighteenth century, cotton fibres were twisted by winding them around a *spindle*, which was driven by the spinning wheel through a combination of a belt and a pulley. The drawing of the fibres was performed by the spinner who pulled them in the direction opposite to that in which the fibres were

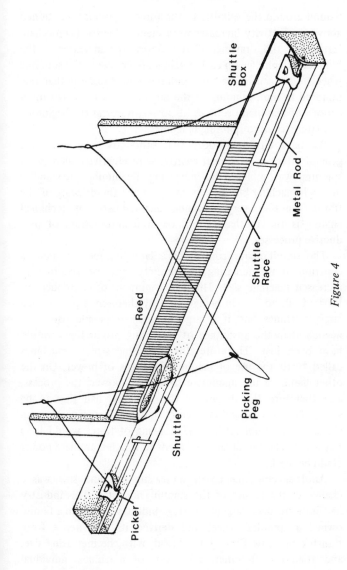

Figure 4

wound around the spindle. In the three inventions mentioned above productivity increases were mainly due to: (*a*) mechanization of the drawing of the fibres; (*b*) increases in the number of spindles on each machine, from one in the spinning wheel to sixteen in the first spinning jenny, to more than one hundred at the beginning of the nineteenth century; (*c*) applications of inanimate sources of power, water at the beginning and later steam, to spinning machines.

We have already noticed that productivity increase is a general effect of technical change. A productivity increase in spinning was achieved by increasing the number of spindles on each machine but also by mechanizing the drawing of the fibres. **Another very common characteristic of technical change is the mechanization, or even automation, of productive processes**.

The simultaneous control of a large number of spindles contributed to increasing productivity, but also required an increased machine size. This had a number of consequences. First of all, no human being could by themselves power such large machines and if a large number of people had been needed then the advantage of increasing productivity would have been lost. Therefore, the increasing size of machines called forth the use of inanimate sources of power. On the other hand, to use inanimate sources of power, the process, and therefore the design of the machine, had first to be simplified and made sufficiently independent from human skills. For example, the hand drawing of fibres would have put a very clear limit on the extent to which inanimate power could be used.

Another consequence of increasing machine size was a change in the nature of the spinning industry. This industry had hitherto been largely a cottage industry, in which a family owned a spinning wheel and derived its subsistence from transforming raw fibres into thread. With the increasing size, and therefore the increasing cost, of machines individual spinners were less and less able to afford them. On the other hand, the few who could afford these machines, with their higher productivity, could make higher profits and, therefore, grow more. Hence the size of the productive units grew and, correspondingly, the buildings in which these new machines were housed had also to be enlarged. Thus, the first textile mills were born. These mills were among the first examples of a new form of organization of production, the *factory system*.

The factory system was indeed an entirely new form of organization of production and not simply an increase in the size of buildings. While previously spinners had owned their means of production (the spinning wheels), in the textile mills of the Industrial Revolution the machines were the property of the owner of the mill, the capitalist. The separation of the property or the means of production from the producers (workers) was perhaps the most important characteristic of the factory system; it was due to a number of factors and not simply to the existence of more expensive and more productive machines.

In order to understand one of these factors let us consider the example of pinmaking. Pins are made out of a metallic wire which has to be drawn, straightened, cut, sharpened; finally a head must be made. All these operations were at one time performed by one worker. But before and during the Industrial Revolution there began a process of subdivision into individual operations of the whole task of making pins. Now one worker drew the wire, another one straightened it, a third cut it, etc., each operation being performed by a different worker. A process of division of labour had begun which was to continue and become one of the features associated with the development of modern capitalism. Adam Smith was one of the first economists to observe this increasing division of labour and to attach to it a fundamental importance in generating economic growth. He estimated that with division of labour the daily output of one worker had been increased from twenty to 4,800 pins. Therefore, very large increases in productivity can be achieved through the division of labour.

The reader will observe that no mention was made in the

previous example of machines. The increased productivity in making pins was achieved simply by subdividing the whole task into more elementary components while using the same tools and materials. Productivity could be increased solely by organizational change, although machines could undoubtedly have given it an extra boost. But were division of labour and mechanization two independent events which contributed separately to increased productivity? Machines, particularly those which existed during the Industrial Revolution, could only perform simple, repetitive tasks. Before a process could be mechanized a division of labour had to take place that transformed it into a sequence of simple, repetitive, unskilled operations; conversely, the invention of a machine could stimulate a further division of labour. This is one example of the absence of sharp demarcation between the clear-cut aspects of technical change and other aspects that might not at first sight be classified as technical change. In general it is probably better to consider process organization and machine technology as complementary aspects of technical change.

The absence of clear demarcation between technical change and organizational, and social, change has at least two consequences. First, it is difficult to give a rigid definition of technical change that applies to more than the most evident cases. Therefore, rather than attempt a precise definition here, I have sought only to examine some of its most important aspects. Secondly, the study of technical change cannot be contained within the boundaries of one discipline but of necessity requires interdisciplinary study, drawing on such fields as the history of science and technology, economic history, economics and sociology.

What are the effects of technical change?

One general effect of technical change has already been mentioned; the increase in productivity due to new techniques and new forms of organization. Through increasing productivity an economy with a constant labour force can produce a larger output. **Increasing productivity, due to technical change, is one of the main sources of economic growth**. The same number of spinners could produce a much larger output using spinning mules than using spinning wheels. If economic growth was the only effect of technical change, everyone would welcome it. Although this is the main justification for technical change, there have been, and there are examples of resistance to its introduction due to other, less pleasant, effects. Some examples drawn from the history of the textile industry will help to clarify these other aspects.

James Hargreaves was forced to leave his native town because of the fear of the local people that his spinning jenny would create unemployment. The city council of Danzig, where the *ribbon loom* (a device which increased the productivity of weaving beyond that achievable with the techniques already existing) had been invented in 1579, suppressed the invention and had the inventor secretly strangled, fearing unemployment among the weavers.

Many other examples of resistance to the introduction of new technology embodied in new machines could be quoted. In general, such resistance arose from the fear that the new technologies would cause unemployment. Were these fears justified? In the case of the textile industry in Britain, we can say that they were not. Between 1806 and 1865, during which time the new machinery gradually but constantly replaced the outdated handlooms and spinning wheels, the total number of people employed in the cotton industry grew from 274,000 in 1806 to 455,000 in 1865. However, although total employment in the cotton industry increased considerably in the nineteenth century, a far more dramatic transformation is represented by the increase in the percentage of factory workers, as opposed to handloom weavers, from 32.8 in 1806 to 99.3 in 1865. The new technology won, but at the cost of very considerable human suffering; afraid to lose their independence, handloom weavers resisted technical change in the cotton industry until, in the end, their wages had declined almost to starvation level.

However, although this was the case in the textile industry during the Industrial Revolution, increasing levels of employment need not be a general consequence of technical change. New machines, which allow higher productivity, can be used to produce either a larger output, in which case the labour force may remain constant or even increase, or a constant or shrinking output, in which case the labour force would be reduced with consequent unemployment. Therefore, **the adoption of new technology is more likely to cause unemployment in a period of economic recession than in a period of high economic growth**. The fears that robots, word processors, etc. may cause unemployment do not appear completely unjustified in an economic recession such as the present one.

As it happens, unemployment has fluctuated very considerably in the last two hundred years, increasing in periods of recession and decreasing in periods of expansion of the economy. Given that, as we have seen, technical change can influence both economic growth and employment we can ask ourselves the question, 'to what extent were these fluctuations in employment and growth due to technical change?' If technical change had always been taking place in the same way all the time there should be no reason for such fluctuations, at least in so far as technical change can have an effect on them. Indeed, economic growth and unemployment are influenced by other factors as well as by technical change and a complete explanation of both phenomena would require consideration of all these factors. Here, however, only the implications for technical change will be dealt with. In particular, two characteristics of technical change will be briefly analysed: the *degree of interrelatedness* and the *degree of novelty* of technological innovations.

Different technologies may be interrelated, for example, in the sense that the output of one is the input of another, and an innovation in one calls forth an innovation in the other. We have already seen how the increase in the efficiency of weaving, due to the introduction of the flying shuttle, stimulated the invention of more efficient spinning machines. The textile mills of the Industrial Revolution soon started using inanimate sources of power, initially water. Clearly the use of water as a source of power placed limits on the quantity of power available and on the location of factories which, in turn, encouraged the adoption of the steam engine as a power source. The steam engine itself had been invented before the decisive advantages in textile technology previously described took place. However, the first steam engines had been invented by Savery and Newcomen to pump water out of mines and were not very suitable for use in the textile industry. The application of the steam engine to the textile industry was facilitated by the invention of the Watt steam engine, which was more efficient and which, by means of the sun and planet gear, produced a smoother rotational motion, more suited to the needs of the industry. More examples of interactions and complementarities between different technologies could be given. However, at this point it is already possible to note that the Industrial Revolution was not caused by one or a few isolated innovations but by a cluster of interrelated innovations. Many of these innovations are also characterized by considerable novelty or superiority with regard to the previous technological practices. **In general, at least since the beginning of the Industrial Revolution, prolonged periods of high economic growth and high employment have been associated with clusters of interrelated innovations of considerable novelty**. The associated effects of all these innovations led to a rate of economic growth higher than any of them individually could have produced.

The industries which during the Industrial Revolution grew much faster than the average and were the leaders of a new industrial order (textiles, coal, steam engines, iron and steel, basic chemicals), subsequently became much less dynamic and lost their supremacy to new industrial leaders. Motor cars, electrical engineering, electronics, plastics, computers and aircraft are examples of industries which emerged in later periods and experienced similar initial periods of very rapid growth. Like their predecessors, these industries increased their share of the economy considerably during this initial period and then went into a period of declining growth or, in some cases, even of stagnation. In this way a continuous process of *restructuring* of the economy takes place in which new actors appear on the scene while the role of the old ones shrinks or disappears.

But is this process of restructuring really continuous or does the rate of restructuring fluctuate in time? This problem is being actively debated today. Among the supporters of the discontinuous character of industrial restructuring are the followers and interpreters of the *long wave theory* according to which economic growth, profitability, investment and unemployment fluctuate in a cyclical or quasi-cyclical manner with a period of fifty years. During the *upswing*, lasting for twenty-five years, economic growth, profitability and investment would be high and unemployment low. This high economic growth would be related to, if not caused by, a cluster of interrelated innovations of considerable novelty. (An example of such a cluster would be the set of innovations in the textile industry which was discussed above, together with associated innovations in steam engines and in chemicals, especially with respect to dyes and bleaching processes.) During the following twenty-five years, the *downswing*, economic growth, profitability and investment would decrease and unemployment would correspondingly increase. Perhaps one of the strengths of the long wave theory is the prediction that periods of high unemployment should follow one another at intervals of approximately fifty years, just as the high unemployment of the 1980s follows that of the 1930s.

Another prediction that can be deduced from the long wave theory is that the present recession will end towards the end of the 1980s. Economic growth, investment and profitability should then recover and unemployment should return to tolerable levels. But where are the technologies which will be the leaders of the new economic order? Are they microelectronics, new energy technologies, biotechnology, information technology? No consensus has been reached in the debate about long waves or about the allegedly discontinuous nature of structural change. Therefore, no answer will be given here to my last question, but it will be left to the reader to explore it.

Questions for discussion

1. What are the effects of technical change?

2. How does technical change contribute to economic growth?

3. What is the effect of technical change on employment?

4. What is the relationship between technical change and structural changes in the economy?

5. Questions which have not been explored here but which the reader could attempt to answer, are:

What is the effect of technical change on power relationships in society?

What is the effect of technical change on work quality?

What is the effect of technical change on the environment?

Further reading

The history of technical change is discussed in books about economic history and about the history of technology. D. S. Landes, *The Unbound Prometheus* (Cambridge University Press, 1969), is a good example of the first approach. A standard reference on the history of technology is *A History of Technology* by Singer, Holmyard and R. Hall (eds.), Oxford University Press, covering the history of technology from ancient times to the twentieth century.

E. Mandel, *Late Capitalism*, New Left Books (1975).

Contains a discussion of the long wave theory.

An invaluable reference on the factory system and on the power relationships between workers and capitalists is Marx's *Capital*, available in many paperback editions.

A book which examines aspects of technical change not dealt with in this chapter, aimed at an audience not having a previous background in the field, is K. Green, C. Morphet, *Research and Technology as Economic Activities*, Butterworths, 1975.

8

CHIPS, BUGS AND SATELLITES: NEW TECHNOLOGIES AND SOCIAL CHANGE

Kenneth Green

This chapter is about the social and economic effects of new technologies; by 'new' I mean not only those technologies based on applications of microelectronics (chips) but also developments in telecommunications (like satellites) and in techniques of manipulation of biological material (bugs) — information technology (or informatics) and biotechnology respectively. There are other technologies, such as those involved in making useful new sources of energy or in making available previously unexploited mineral reserves or in conserving and recycling expensive raw materials, but they will not be referred to in this chapter.

Writing about new technology has mushroomed. Over the last three years at least twenty popular or introductory accounts of new technology and its supposed 'effects' on employment, industrial structure, skills, living patterns, educational requirements, etc. have gone on sale in Britain. ('Effects' and 'impact' will usually be in inverted commas for reasons to be outlined later). The fact that the current slump is hitting Britain harder than any other developed capitalist country and that therefore British discussions of new technology (particularly its employment 'impact') have tended to take on an apocalyptic air should not lead us to think that the British are alone in their concern. The International Labour Office (ILO) publishes a quarterly abstract service of developments and publications relating to labour matters. For the last four years the ILO has summarized reports from all developed capitalist, many state socialist and even some developing countries on what new technologies — particularly those based on microelectronics — might mean for their economies. One might conclude that interest in technologies and their social 'impact' is enjoying a popular revival — not merely among professional soothsayers but among state and industrial decision-makers as well as trade unionists and people in the street. The interest is worldwide and is reflected in the growth in circulation of science magazines and popular computing and electronic journals.

There is insufficient space here adequately to enter the debate over precisely what new technologies can offer society and what their 'impacts' will be. Such a discussion would only add yet another version of what new technologies are all about to an already overlong list. Anyway, I would like to avoid the mere chronicling of who says what, of how many unemployed by 1995 are predicted by X and Y, and which is right. The bibliography will give rapid access to the technical details of computer and biological technologies as well as to the debates over their impact, if you require them. Instead, I want to offer a perspective on the social and economic 'effects' of new technologies which will, I hope, be a means of bringing into the open the political perspectives and human values which lie at the centre of the debate over any fore-casting of the future.

Nevertheless, it would be useful to remind the reader of the essential features of those new technologies about which concern has been expressed. The first of these is the 'micro-processor' — the microchip — essentially a tiny computer which can be incorporated into almost any machine or device permitting an extension of the machine's capacities or even giving rise to machines which can perform completely new tasks. The use of computer-controlled robots (now, thanks to microelectronics, much cheaper and more versatile), of computer-controlled flexible metalworking machines and of cheap computer-controlled chemical processes allows rapid rises in productivity. This can lead to a corresponding drop in the number of human machine operators — towards the 'workerless factory' — or to a deskilling of those who remain. When incorporated into domestic consumer products, microelectronic components can either extend the range of existing machines (computerized washing machines, electronic controls for hi-fi units) or can give birth to a new set of machines otherwise impossible to make at the right size, price and ease of use (like TV games or home computers). In combination with other technologies – satellite relay stations, optical fibre transmission cables, electronic storage and switching systems – microcomputers permit the 'electronicization' of all communications which, up to now, have been carried out by means of a variety of very labour-intensive media: writing, typing, typesetting, mail carrying, and so on. With the use of word processors, storage discs, video screens, electronic telephone networks, it is now possible to manipulate words, graphics and data – in short, information – and collect, transmit and present them much more rapidly and cheaply than before with a lot fewer people. This potential 'information revolution' has implications for a whole variety of non-manufacturing industries and jobs.

Its potential application is not just to service, white-collar industries but to professional activities (computerized medical diagnosis, computerized teaching machines) and to the conceptual side of manufacturing (computer-aided design and the electronic coupling of design, production and co-ordination into computer-aided management) as well. The potential applications of the microchip are pervasive, and this explains why silicon chip technology is seen not merely as an extension of previous computer technologies but as a quantum jump – a heartland technology of even more significance than such equally pervasive but more familiar technologies as the electric motor or plastics.

New technologies based on recent developments in the biological sciences could have equally important implications for economic life and social arrangements. Already, genetically-engineered bacteria are being used to produce small quantities of very expensive materials for the treatment of hitherto difficult-to-control medical conditions; interferon and insulin are being made this way. The prospect is that the range of drugs that the pharmaceutical industry will be offering by the end of the century will be much larger, and that completely new types of therapy will have been developed based on them. Genetic manipulation and the cloning of plant and animal species could have an even more profound socio-economic effect. Food processing could be considerably changed and drastically reformed, by using crop strains which give higher yields in inhospitable soils and climates; or we could contemplate the completely synthetic production of basic foodstuffs by developments in fermentation (a long established biotechnology), such as the use of engineered algae to make milk proteins from plants, thus eliminating the need for dairy farming. Other potential areas of application for biotechnologies include the chemical industries (production of industrial chemicals by new systems of fermentation), mineral extraction (using enzymes in huge quantities to digest copper and uranium), and pollution control (using engineered bugs to scavenge, and thus cleanse, otherwise polluted water and air emissions).

So, just as one might extrapolate from robots and computers to workerless factories, with product requests going in at one end and the finished articles emerging at the other, virtually untouched (and indeed undesigned and unmanaged) by human hand and mind, one could, equally, extrapolate from genetic engineering and cloning techniques to envisage food and pharmaceutical factories growing raw materials from basic chemicals and processing them into drugs, beans or simulated chicken, a full biochemical integration of agriculture and food processing – the automatic, workerless, biotechnological factory!

Popular presentations of the applications of these new technologies tend to emphasize the long-term dramatic impacts that can be envisaged – hence the microelectronics *revolution*, the information *revolution*, the biotechnology *revolution*. But since hindsight is the only exact science, any attempt to predict the long term 'impact' of a particular technology is inherently hazardous. This is especially so with technologies which possess wide applicability and which have so many *potential* interconnections with all social activities. However, popular conceptions of the microelectronic, information and biotechnological revolutions suffer from a number of failings. Three will be examined: technological optimism, lack of appreciation of the significance of the current economic depression, and insensitivity to national and regional variations in a technology's 'impact'.

Technological optimism

There can be no denying the *potential* of new technology to change drastically industrial structure and living patterns. But the realization of this change is bound to take some time, the exact length of which will differ depending upon the applications in question. To determine this time-scale, it is insufficient to know in principle how, say, some microelectronic component might be used to make a new product or allow a radical reorganization of an industry. The products and machines for such purposes have to be *designed*, the computer programmes to make the chip work need to be written, the connection between the new technology and the existing production systems in which it will have to fit need to be worked out, the resulting system will have to be de-bugged and, finally, customers will have to be found. Each of these tasks requires considerable research and technological development and, therefore, time. Despite popular images, computerized robots reliable enough to be trusted to keep going in industrial environments are really quite limited in their application — principally to extremely simple tasks in mass production industries. (The Fiat Strada, contrary to its advertising propaganda, is not *built* by robots; its body shell is *welded* by robots, and then not completely — its assembly is still accomplished by living labour.)

Economic depression

Hardly any of the unemployment currently being suffered in Britain can be attributed directly to the use of new technologies. Although the application of microelectronics to certain industrial products by firms in other countries has intensified competition in those industries, sometimes to the detriment of British firms which have not been so technologically dynamic, the current slump can be explained more by reference to longer-term structural features of the entire world economy than to labour-displacing technical change. Slumps are an essential part of capitalist economies; they have a function which bears upon technological change. During slumps, competition for reduced markets increases, thus weeding out weaker firms and promoting rationalization and industrial concentration. This process inevitably implies the shedding of labour when profit is the principal yardstick of economic success. Such rationalization may involve no more than increasing the efficiency of existing machinery and systems of production. More often, however, it involves the introduction of new equipment, a change in the products being made, or a redistribution of productive activity within the company either within one country or, in the case of multinationals, around the globe. It is difficult therefore to separate out, say, the employment impact of new technology because it is part of a complex economic process of industrial restructuring that is taking place anyway. The chain of causation — from particular pieces of machinery based on some new technological development through their introduction into particular industries, to their effects on productivity, manning levels and, thence to redundancies and unemployment — is difficult to unravel, particularly if the chains of several industries are cross-linked, as they are in a complex economy such as Britain's.

Regional differences

Clearly, because industries and offices are located in specific places in particular countries, the impact of new technologies will differ between countries and between regions within one country. Inevitably, then, the resulting changes in employment or skills required will be differentially distributed. An over-emphasis on the national level − forecasting unemployment for the UK as a whole for example − will miss the regional variation. Yet it is at this level that people actually perceive the effects of many industrial changes and it is those percep-tions which influence their responses to those changes.

Without narrowly prescribing the technologies concerned, the time period being considered, the range of impact to be looked at and even the geographical regions concerned, any prediction about a new technology's 'impact' can only be a generalized speculation. Therefore, given the enormously large field of applications for micro- and biotechnologies, studies which aim to identify *significant* impacts are bound to vary. Here are a few views about the degree of unemploy-ment likely to be caused by microelectronics in Britain, culled from various popular discussions of new technology over the last few years.

'Six million unemployed in the UK by the end of the century is not unlikely' (T. Stonier, Professor of Science and Society, University of Bradford).

'It is likely that the miseries and upheavals of the 1930s will pale into comparative insignificance should Britain con-tinue to drift on as at present, and find itself in the situation where 2, 4, 6, or even 7 million are out of work' (Colin Hines, Earth Resources Research).

'Three to four million jobs could be destroyed by the microchip revolution' (Ian Lloyd, Conservative MP).

But, there are other, more optimistic views:

'It is important that we appreciate the scope of robots in the industrial environment. A proper approach to robot implemen-tation not only improves the productivity of conventional and on-going manufacture but *sires new industries to stimulate employment* and wealth generation' (G. L. Simons, National Computing Centre (emphasis added).

'It seems that microelectronics is likely to create new opportunities for work. Scientists, engineers and technicians will be needed to design and manufacture the new devices, and people of many levels of skill and training to use and apply new machinery in new work situations. The hope is that it is not rewarding work that be reduced, but drudgery and unnecessary hard, undignified or disabling labour' (W. H. Mayall, Science Museum).

Mayall's optimistic view about the survival of rewarding work is contradicted by statements like this:

'Word processing also has some unwelcome possibilities. Typing of letters and reports is already often organised, in "typing pools" in a manner which makes it resemble factory work of a routine kind. The word processor could allow this organisation to be carried a stage further, with less skilled typists "entering information" through keyboards with no means of checking or correcting errors, but with the highest possible speed. Correcting errors would then become a specialised occupation, carried out at specialised terminals. The job of the secretary, in the past, has been a varied one, often including a wide knowledge and responsibility. But if a secretary's effectiveness in typing falls much below that of a word processing system, that job may also be fragmented, as it already is in some places: typing and filing is removed to the central pool, and other services such as diary and answering the telephone are provided on the basis of one secretary to six or eight managers' (Council for Science and Society).

The wide variety of views on the impact on new technologies on office employment is summarized in Table 2.

Away from employment, consider two views about the threats to personal privacy that elaborate computerized record systems often exhibit:

'The new technology of microelectronics has proved a useful tool for the organs of state security. Microelectronics makes extensive surveillance through 'phone tapping and bugging a much easier task. It also extends the power of state security forces to record and process information on the

Table 2   **Forecasts of impact of new office technologies on office employment**

(Adapted from E. Bird, *Information technology in the office: the impact on womens' jobs*, Equal Opportunities Commission, 1980.)

Report (Country, date of report)	*Type of job affected*	*Effect*
Siemens (West Germany, 1978)	Office in West Germany	By 1990, two million typing/secretarial jobs lost (40% of total office employment)
Nora/Minc (France, 1978)	Banking/Insurance	30% reduction by late 1980s
Association of Professional Executive, Clerical and Computer Staff (Apex), (UK, 1979)	Typists, secretaries, clerical, document authors	250,000 jobs lost by 1983
Association of Scientific Technical and Managerial Staffs (ASTMS) (UK 1979)	Information processing jobs	30% loss by 1990
Conservative Party (UK, 1979)	Private sector clerical and administration – insurance, bankers, building societies	40% of jobs at risk in 1980s
Department of Employment (UK, 1979)	Banking	Modest *increase* in employment till 1985; some decline in 1990s
	Insurance – clerical data preparation jobs	15% decline by 1985 unless new services offered
Equal Opportunities Commission (UK, 1981)	Typist/secretarial	20,000 jobs lost (2%) by 1985; 170,000 jobs lost (17%) by 1990
	Clerical	More lost jobs than for typing/secretarial

population. And, finally, it provides much more effective and secure communication between operational units, which increases the effect that a given number of troops and police can have' (Microelectronics Group, Conference of Socialist Economists).

On the other hand:

'The spread of cheap, universal computer power will result in a gradual loosening of restraints on the movement of information within a society. The world of the 1980s and '90s will be dominated not only by cheap electronic data processing, but also by virtually infinite electronic data transmission. With thousands of communication satellites likely to be scattered into orbit by the space shuttle in the next ten years or so, person-to-person radio communication is going to be commonplace in the Westernised world, and global TV transmissions will also become widespread. This kind of development will encourage lateral communication – the spread of information from human being to human being across the base of the social pyramid. Characteristically this favours the kind of open society which most of us in the Western world enjoy today and has just the opposite affect on autocracies – both right and left wing – who like to make sure that all information is handed very firmly downwards' (Christopher Evans, Psychologist and Computer Scientist).

Disagreements over the future for biotechnologies can be equally sharp. Thus:

'For a while (after some of the earliest genetic engineering experiments in the seventies), visions of Frankenstein monsters, or strange viruses gone errant, prevailed. Scientists had learned to create new forms of life. What, we wondered, might happen to us as a result of this awesome new power? But in the past several years time has begun to tell a constructive tale. Instead of Frankenstein monsters we've gotten insulin. And human growth hormone. And, most recently, cloned interferon – the body's own wonder drug. Because of the ease with which genes can be manipulated, or 'engineered', scientists have begun to envision a world in which there's no cancer, no genetic disease, and no birth defects; a world without starvation and with enough energy for everyone's needs. A world in which the average age will be 100' (*The Techno-Peasant Survival Manual*).

Alternatively:

'There are going to be many disappointments. The market
forecasts (for products produced by biotechnologies) are all
guesses. To turn the base metal of biology into big profits
would need not only a lot more basic research but also a lot
more practical experience, a lot of process engineering and
also much bigger investments than people are contemplating
today. Risks will be high, patents hard to enforce, competition
frenetic and most products (when they do come) rapidly
obsolescent' (*The Economist*).

Such diversity of views is hardly novel. Much the same
diversity can be had from positions taken on the construction
of civil nuclear power stations — differing assessments of the
significance of safety factors, of short-term major hazards and
long-term injuries to health, of the costs of construction, of
the usefulness of nuclear power to Third World development
plans, of the dangers of reliance on such skill-intensive high
technologies, of the problems of the disposal of nuclear waste
and its policing for hundreds of years, of the requirements for
more electricity in the twenty-first century. The current
arguments over microcomputers and their social 'effects' can
even be seen as a repeat of similar disputes in the 1950s and
early 1960s regarding the benefits and disbenefits of 'automation' in which a similar range of views was expressed as
today.

It is hardly necessary to point out that some of the present
forecasts about the future with new technology stem from
different, if not from completely opposing, views about how
societies work and change and about what goals those societies
should be aiming for. Although one can sometimes point to
deficiencies of argument and logic in such forecasts by drawing on historical examples of previous technological revolutions, there remains beyond this an irreducible core, rep-
resenting political and human perspectives, which lies at the
heart of all forecasts. I refer to the assumptions which forecasters must make if they are to order the facts and select
significant trends. It is important that, in works which try to

predict new technology's 'impacts', such assumptions be made clear. The first question to be asked, on reading any predictions of a technology's social 'effects' is: what form of society, that is, what set of social and human relationships, does the writer seem to take for granted? Indeed, has the writer *any* sense of social change (as opposed to merely technical change), which is more than just an extrapolation of the present social arrangements? This requirement can best be illustrated with reference to science fiction writing. Many science fiction stories set in the far future and based upon highly imaginative, if somewhat magical, technologies (time warps, matter transfer and the like), display a dismal lack of imagination when it comes to how these societies are to be organized. Emperors, lords and slaves abound and science fiction writers seeking alternative social arrangements dig back into human history's rubbish heap of oppressive social relationships — slavery, feudalism and imperialism boldly go where none have gone before. (The novels of Ursula LeGuin — in particular *The Word for World is Forest* and *The Dispossessed* — satirize this fascination for anachronism by presenting social arrangements based on the longings of currently oppressed people — particularly those oppressed by neo-colonialism and patriarchy.)

A brief classification of the assumptions about social development and political perspective held by writers on new technology's social 'effects' might be helpful in guiding the reader through the burgeoning literature. The classification is only intended to be suggestive; many of the writers on new technology would fit into more than one category — a product of the imprecision of the categories as well as of their own confusions. There are four principal positions — the **free marketeers**, the **state interventionists** (or corporatists), the **radical politicals** and, emerging from the movements of the 1960s and 1970s that criticized the way scientific developments have been applied over the last forty years, the **critical technologists**.

The **free marketeers** (most popularly represented in the pages of the weekly *The Economist*) are positively jubilant at the arrival of new micro-based machines and information technologies. Large corporations, they argue, will no longer be able to monopolize manufacturing production. Up to now, mass production of cheap complex articles has required enormous capital outlay to provide factories full of large, relatively inflexible machines (like assembly lines or chemical process plants). As machinery based on computer technologies becomes cheaper and more flexible, then the opportunities for small firms to burst into those markets previously dominated by large ones will be considerably increased. The utopia of free competition will dawn, as soon as governments relax regulations and create the appropriate economic climate for these thousands of smallish, high-technology, highly competitive companies to boom. Additionally, the revolution in telecommunications will stimulate decentralization and, by allowing massively increased lateral communication, unfiltered and uncensored by TV companies, post office corporations, the state, and big private corporations, will force large bureaucracies to disband. Small will indeed be beautiful, based on microelectronics.

The **state interventionists** (or corporatists) are far less sanguine. They recognize that free-market dreams are based upon a view of social organization that ignores politics; that is, it ignores the obvious fact that in the absence of some rigid authoritarianism, large sectors of the population will not be content to subject themselves to the whims of a free market. Highly organized, urbanized, technology-dependent, economically unequal, continually changing societies need 'managing' and planning. The setting of wage levels, educational provision and training, ironing out regional disparities, etc., all require state involvement and, in consequence, a fair degree of consensus between the various classes and interest groups; this in turn depends on some degree of social stability and avoidance of economic depression. Trade unions, which in all capitalist societies seem to act as corporatists, argue for technology agreements and negotiations over the introduction of machines and processes; large corporations, state or private, seek government support for research, for cheap raw materials, for regional subsidies and the like. Governments attempt to manage industrial relations, provide venture capital for potentially strategic new industries or block imports of foreign goods by tariffs, quotas and special regulations. Of course, different countries manage these matters in different ways but they generally believe that such management by the state is essential to control the adverse effects of new technologies.

The **radical political**'s response is to deny that there is anything automatic about the free market or corporatist management of capitalist economies in their ability to guarantee living standards and preserve meaningful work in present economic conditions. The normal method of dealing with the introduction of radical new technologies is by dramatic restructuring of the whole economy, and indeed the whole society. Through economic depressions, with all their implications for redundancies, deskilling, regional decline and reductions in welfare service, firms are closed and new industries are started up, governed not by social need but by the requirements of profit making and competition. Through corporatism the state tries to manage the results of the 'anarchy of capitalist production' but does not get to its roots. The initiative of private firms and their profit seeking is left unchallenged; in fact, corporatist governments diffuse opposition to capitalist rule by restricting trade union struggles and manipulating public opinion.

The **critical technologist**'s response accepts much of the radical political case but criticizes it for its 'reactive' nature. Although socialist activists and leftist trade unionists seek to organize opposition to state and capitalist strategies, they delay the debate on *technological* alternatives (as opposed to alternative forms of social organization) to some future society where, it is supposed, such alternatives would become practically possible. The critical technologist would see the future not as the product of some autonomous and, indeed, inherently progressive, technological force but as being actually *created* – by social groups contesting with each other over costs and benefits (economic, social and moral) of proposed changes. Such an understanding is concealed by constant reference to new technology's 'impact' on society, the economy or whatever (and I have used inverted commas around 'impact' or 'effects' for this reason). Technology does not exist (or is not created) outside the social process, to be propelled into 'society' – into factories, the market place, offices, military installations and so on, with resultant 'effects' on industrial relations, firms' sales and profits, peoples' living standards, etc. New technologies are themselves imagined, researched, developed, introduced and put into operation – in short, they originate – over a prolonged period within specific social arrangements with certain initial aims; these may be to increase sales, to reduce labour costs, to eliminate some technical bottleneck, to weaken the power of some sections of the workforce, to drive a competitor from the market, to gain military superiority, to facilitate surveillance of subversives, or to keep law and order. In this process of origination they will be subject to counter-pressures either from the state (by means, for example, of safety legislation for workers and purchasers), from organized workers (refusals to work with certain technologies without changes in payment levels and training arrangements), from unorganized workers (passive and active sabotage of managers' and engineers' plans) or from 'issue' pressure groups (residents' groups opposed to a new factory in their area, women's organizations opposed to some new product or drug regarded as perpetuating women's subordinate status). In short, a variety of futures in which new technologies' potential is realized in all sorts of ways is possible, depending on power relations between social groups. This in turn, depends, at least partly, on the groups' knowledge of what is possible and on an appreciation of the view that there is no *inevitable* direction of technological development and no inevitable social impact.

If you accept the view that the new technologies of micro-electronics, informatics and biotechnology are likely to be implicated in dramatic social, economic and political change but that none of the details of that change are inevitable, then the conflicting opinions and forecasts over likely employment levels, the skills that will or will not be in demand and so on, cease to be confusing. They offer, to a greater or lesser extent 'visions' of the future which has yet to be *made*, by political action in its broadest sense.

Questions for discussion

1. Teaching on the topic of technology's social 'effects' is of limited usefulness if it does not move quickly into questions of *policy*. What can the members of political parties, trade unions, professional associations, local councils, national government bodies and committees, women's organizations and pressure groups of all sorts *propose*, and how should they *act* to guide technological developments towards progressive social goals?

2. What criteria for judging the worth of proposed new technologies could you list?

3. What information might you need to decide between the views (quoted above) of (i) Mayall and the Council for Science and Society on quality of work, (ii) Conference of Socialist Economists and Evans on personal privacy?

4. How would you respond to the argument that it is essential for British workers to accept new technologies, otherwise their foreign competitors, who do accept them, will become more efficient and will take away British companies' markets and thus put even more British workers out of work?

Further reading

T. Forester (ed.), *The Microelectronics Revolution: The Complete Guide to the New Technology and its Impact on Society* (Basil Blackwell, 1980).

A comprehensive collection of 41 readings on micro-electronic-based technologies covering: the technology itself, the chip-making industry, applications of the technology, the economic and social impact in manufacturing and office work, employment and industrial relations implications, broader social impact of information technology and the 'information society'. There are pictures, diagrams and extensive guides to further reading. The full range of positions described in this chapter – free marketeers, state interventionists, radical politicals, critical technologists – are represented in the readings. (See if you can spot them!)

Cabinet Office, Advisory Council for Applied Research and Development. *The Applications of Semiconductor Technology; Joining and Assembly: The Impact of Robots and Automation; Computer Aided Design and Manufacture; Information Technology; Biotechnology* (HMSO, 1978, 1979, 1980).

A series of short pamphlets written by a committee of industrialists (with a few senior academics, civil servants and trade unionists) which advises the Cabinet on policies for the promotion of technology. The pamphlets describe the basics of each technology and make recommendations for government departments and quangos on appropriate technological strategies for state support, education and training policies, required changes in the law, etc.

A.M. Cunningham and five others ('The Print Project'), *The Techno-Peasant Survival Manual* (Bantam Books, 1980). (Techno-peasant: anyone who is technologically illiterate; a person whose future is in the hands of the technocrats.)

Subtitled, 'The book that de-mystifies the technology of the 80s', this is a mind-blowing survey of new technologies: microcomputers, fibre optics and lasers, genetic engineering, space shuttles and satellites, fusion power, weapons technologies, artificial intelligence. Excitingly presented with lots of graphics, it often goes over the top in its attempt to instruct the technologically illiterate on what is in store — but, unlike most books on new technologies, it is a good and provocative read.

Council for Science and Society Working Party, *New Technology: Society, Employment and Skill* (CSS, 1981). (Available from 3/4 St Andrews Hill, London EC4 5BY.)

This report throws doubt on the view that new micro-based technologies add up to a 'revolution'; instead, it sees possible effects on employment and skills as being just new expressions of the continuing trend to displace some workers and deskill others, in pursuit of supposed immediate benefits but ignoring the longer term disbenefits of unsatisfying work and social alienation. The account of the development of machines, automation and the division of labour and the discussion of the need for developing technologies that *enhance*, rather than displace, skill for all workers is worth reading.

J. Rada, *The Impact of Micro-electronics: a Tentative Appraisal of Information Technology* (International Labour Office, 1980).

One of the few books about new technology that sets any developments in an *international* context, examining the global spread of computer technologies, the influence of capitalist competition and effects on industrial structure and employment prospects in Third World countries.

Conference of Socialist Economists, *Microelectronics: Capitalist Technology and the Working Class* (CSE Books, 1980).

A socialist analysis of the problems new technology poses for working-class organizations. Includes analyses of capitalist strategies for new technology in a variety of industries (office work, process, small batch engineering, motor vehicles, mining, banking) and of the responses of the capitalist state in the spheres of education and training. A chapter on 'Alternative design' suggests how alternative, socialist policies for new technology might be envisaged. A hybrid of radical political and critical technologist types of response.

E. Yoxen, *The Gene Business: Who should control Biotechnology?* (Pan, 1983).

This book is an account of the development and promise of biotechnology, scientific, technical, industrial, cultural and political. As the sub-title suggests it asks who (scientists? industrialists? the public?) should choose in which directions biotechnology should go.

9

ENERGY: OPTIONS, PROBLEMS, POLITICS

Roger Williams

'Energy' remains among the more straightforward concepts in physics, technically denoting the capacity of a system for doing mechanical work, but since 1973 the word has also taken on a much wider meaning. 1973 was, of course, the year of the first world oil crisis, and 'energy' now stands above all for the total inventory of fuel and power supplies available, or conceivable, for meeting global demands, the latter themselves depending to a greater or lesser degree on the types and prices of the various supply sources on offer. Similarly, 'energy policy' has now become a shorthand label for the delineation by governments of tactical and strategic objectives in this field, together with the development of measures for pursuing them, a simple enough idea in principle, but in practice a highly complex and far from certain undertaking.

Dimensions of energy policy

Energy policy is necessarily the resultant of many very different, though usually interrelated, factors. Let us begin with scientific and technical considerations. One common problem here is that the scientific principles behind some process are well enough understood but no method has as yet been found or proved for embodying these principles in a technology capable of safe, reliable and cost-effective operation. Probably the best example here is thermonuclear fusion. This has long been known to be the energy source of the stars and scientists have a sound grasp of its fundamentals. But although it proved possible thirty years ago to utilize the energy produced by the fusion of light atoms in an uncontrolled and highly destructive fashion − in the so-called hydrogen bomb − despite a very large, worldwide and co-operative research and development effort, a viable device for obtaining useful energy from the process still seems decades away. Indeed, it cannot be said with confidence that such a device will ever be constructed. Another important example of the same basic difficulty concerns the various arrangements proposed for extracting the energy in ocean waves. Here as with fusion, initial hopes have given way to harsh realities. A somewhat different technical problem is encountered in the case of the fast-breeder nuclear fission reactor. This has always been viewed by advocates of nuclear power as the apotheosis of fission technology in that it is able to consume the plutonium produced as a by-product in the first generation of electricity-producing nuclear reactors, thereby making possible in theory a fifty-fold increase in the utilization of the energy in the original uranium. The question mark in this instance lies not so much in the feasibility of the technology as in its safety, and therefore suitability for commercial operation. In still other cases where the problem is essentially technical, the basic technology is fully proved and safe, and may even have long been in use, but a need is perceived for a larger scale of operation or for the achievement of a higher efficiency up to some theoretical maximum.

At what can be considered the opposite end of the spectrum from the scientific and technological issues in energy policy are the political ones. Here one is concerned with how exactly public decisions are made in a particular society, and therefore with that society's institutional arrangements, degree of centralization, and so on. As with technical difficulties, the political problems which can arise are of many different kinds. To begin with, there are both international and intra-national issues. The best example of international disagreements are those which occur between the main oil-producing and oil-consuming countries, further disagreements arising within these two groups themselves. Within a state, arguments must be anticipated between the main bodies charged with energy supply, the coal, oil, gas and electric power authorities. Employees in these industries will from time to time feel disadvantaged or threatened and in consequence bring pressure to bear. Consumers will have grievances and, increasingly, objections must be expected from sections of the public when new energy facilities are proposed, whether these are coal mines, nuclear power stations, gas storage tanks, or whatever. Impinging on the political dimension of energy will be social and psychological considerations. Thus, high energy prices will bear especially heavily on the more vulnerable members of society, raising the questions of supply disconnection and subsidy. And the behavioural disposition of consumers and of the public cannot be neglected, in particular for example, in evaluating the possibilities for energy conservation and in balancing the risks as between different energy sources. This latter takes one into an area sometimes described as social engineering (when its objectives are approved by the observer in question, as usually in the case of conservation), or as unscrupulous manipulation of public opinion (when they are not, as commonly with public relations exercises in furtherance of nuclear power).

Sandwiched between the technical and the political aspects of energy policy are the economic ones. The central focus here is the effect of prices on supply and demand patterns, which in turn leads to attempts to co-ordinate the movement of supply and demand over time. It is doubtful whether the frequency of truly free markets is any greater than that of free lunches, but in the energy field, as in others, some markets are certainly more free than others. Governments normally pay close attention to energy prices, now yielding to the urgings of producers, now responding to the complaints of consumers, all the while with one eye to the international situation and another to their own ideological and electoral interests. To speak of an 'energy gap' is unhelpful but prices will obviously affect the manner and level at which a balance between supply and demand is finally struck, as well as subsequent public and private initiatives to change this (via technological innovation, the discovery of new reserves, and altered consumer behaviour). In the energy sector, as more generally, economics aims to be scientific but it is well to remember that it remains a highly constrained science, the constraints deriving both from inadequate precise information about the existing situation and from the limited possibilities for predicting quasi-economic events (such as the level of industrial activity and changes in energy use efficiency) and non-economic ones (such as political revolutions and climate fluctuations). Energy departments have no choice but to forecast. They must have some model of the future against which to test current options. It should come as no surprise that the 'energy crisis' years after 1973 have seen a great expansion in forward projections: it is possible to be cavalier about the future only if one is confident it will be benevolent. Unfortunately, the past exerts a major influence, both in the real world and in the minds of forecasters, who may also find themselves unconsciously elevating their working assumptions into certainties. The computing power available today has perhaps provided a torch where formerly there was only a candle, but the night remains dark and the wise forecaster's watchword is still humility. What forecasting can do is to throw some light on relationships within the energy economy, bring out the connections between this and the wider economy, identify limits, suggest options and in general upgrade the quality of discussion about possible energy paths. Larger claims can safely be ignored.

To the political, economic and technical dimensions of energy policy must be added still wider influences. Meteorological considerations evidently have a bearing both on demand possibilities and on certain supply possibilities, and the latter are also very substantially shaped by geographical and geological facts. It is also salutary to reflect on the momentum generated by immediate past energy policies. Thus, the era of cheap energy ended in the 1970s, but because that era effectively discouraged the investigation of alternative energy sources such as solar, wind, wave and biomass energy, any substantial contribution from these sources must now await the outcome of lengthy research and development programmes. As an equally important example, the initial development of nuclear energy took place in a far more benevolent public environment than that which came to prevail worldwide in the 1970s, yet the inertia formed by the earlier nuclear commitment and its institutional embodiment has continued to exert a powerful influence on governmental thinking.

Energy supply considerations

So much for the main elements of energy policy: what are some of the key features of the principal forms of energy supply, coal, oil and nuclear power? The recent history of the coal industry internationally offers an excellent, and cautionary, demonstration of governmental short-sightedness. In the sixties, coal was regarded very widely as an energy source of the past, whose contribution must inevitably decline. In line with this, the coal industry was savagely reduced in a number of countries, most notably in Japan, but also in the United Kingdom. In the seventies, in the context of uncertain and expensive oil and growing public opposition to nuclear power, the general assessment of coal's role changed dramatically and it is now expected to be an important industry of international commerce more or less indefinitely. There remain major environmental reservations so far as newly affected communities are concerned – the Vale of Belvoir in England is a good case in point – but these environmental reservations still tend to be viewed as qualitatively different from those attaching to nuclear power. Meanwhile, many new research and development projects have been put in hand, many of them involving international cooperation, and there is active interest in new coal-burning technologies, in liquifaction possibilities to produce synthetic oils, and in advanced processes of coal gasification.

Nuclear power gives rise to problems both substantively and in respect of its public acceptability. The principal substantive problems arise in respect of the safety of the various facilities, in particular power reactors, spent fuel-reprocessing plants, and final waste-disposal sites. Also of major concern is the relationship between nuclear energy development and nuclear weapons proliferation – nuclear energy development is not the most attractive route for a state seeking nuclear weapons, but it is undeniably one such route. Safety, as usual, is closely related to cost, in that engineering further safeguards into a plant naturally tends to increase its costs, disadvantaging nuclear power in comparison with alternatives. As to the public acceptability of nuclear power, significant groups, if not the public at large, tend to be disturbed about the adequacy and implications of the solutions adopted to most of the substantive problems. They question the safety of nuclear plant, both as regards permitted environmental discharges of, and occupational exposures to, radiation and with respect to the assumptions made about accident probabilities and consequences. They argue that the security arrangements necessary to protect nuclear plant from criminals and terrorists must involve a serious general loss of civil liberties. And they are reluctant to accept that the development of nuclear energy is other than central to the acquisition of nuclear weapons. The public acceptability aspect of nuclear power, given its roots in genuine substantive issues, came in the decade of the 1970s to constitute a major check on nuclear development, and legal intervention, political action and disruptive civil protest by those opposed to nuclear power, gave governments and utilities occasion to at least pause in their commitment to the technology. Halts or delays were especially marked in Austria, Sweden, West Germany and the USA, and only highly centralized governments like those of France and the USSR were able to keep closely to their initial plans. Given that nuclear technology is no more capable than any other technology of remaining indefinitely accident free, it seems likely that its future depends critically upon the distribution, severity and timing of those incidents which do occur. Thus, the Harrisburg accident in the USA in 1979 had a marked international effect, yet in the event, despite real fears at the time, this posed only a small radiation hazard, though it was a financial disaster of major proportions. An accident involving a serious discharge would be a quite different proposition, especially if it occurred while nuclear energy was still establishing itself and in one of the more advanced industrial countries.

While worldwide nuclear power has become the energy technology which most impacts on national politics, the case of oil is even more striking in respect of international politics. The high proportion of the world's oil reserves located in the politically unstable Middle East makes security of supply the most worrying problem, and to differing extents the western industrial nations have allowed themselves to become dangerously vulnerable to supply interruptions both calculated and incidental. The success of OPEC in bringing together most of the key producing countries, and the determination of individual states to get the most that they can for their oil, between them have led to a second major problem centring on price. With predictability largely gone, although price fluctuations will occur, the long-term trend in prices would seem unmistakeable. High prices in turn mean that poor countries are further disadvantaged. They also produce a distinct shift in international economic power, the rapid changes in this respect in the seventies having undoubtedly contributed to international economic recession. And domestic oil and gas production themselves are capable of distorting a state's economy, an awkward problem, despite the positive advantages, in cases like those of the Netherlands and the United Kingdom where the manufacturing sector has tended to suffer from a high, because oil- and gas-related, exchange rate. The severe difficulties precipitated by the first oil crisis (1973) forced the western nations into at least a measure of cooperation through the creation of the International Energy Agency. It also led many governments to establish departments specifically concerned with energy. Subsequently, many countries placed increased emphasis on nuclear power as a substitute for oil, a shift, as explained above, with large political ramifications. They also redoubled their efforts to discover and bring in new sources of oil, a development with its own wide implications. At the same time, the formulation of oil-depletion policies has become a central matter for all oil-producing countries.

The possibility was referred to above of extracting the energy in ocean waves. This is one of several examples of what have come to be known as renewable energy sources, to distinguish them from fossil and nuclear energy sources where the energy is derived from a fixed original stock. Some 'renewable' sources are well-proven, in particular tidal and hydro-electric power, and most of the others, solar, geothermal, wind and biomass, have some demonstrated success on a limited scale. Currently a major international effort is in progress to determine the true potential of these various renewable sources. Different countries are emphasizing tech-nologies likely to be of particular interest to themselves — thus Britain and Japan as islands initially concentrated on the possibilities of wave power — and only by the end of this century can one expect a true picture to emerge of the ulti-mate contribution to be expected from 'the renewables'. As well as the technological problems general to them, to be widely adopted they must also be cost effective, and with some of them, solar power in particular, the balance remains to be determined between small-scale decentralized systems and large-scale centralized ones. Currently, many of the 'renewables' are still at the research and development stage where costs are relatively low, and governmental benevolence to them may change when demonstration projects begin to absorb substantial finance. The renewables collectively have been embraced by those opposed to nuclear energy but it should not be overlooked that they, too, will have their environmental impact.

Given that there are difficulties and risks, both physical and political, with all energy-supply technologies, it has become natural to consider the possibilities for reducing and modifying energy demands. This approach has come to be thought of as energy conservation, though energy efficiency would really be a better term. While energy was plentiful and therefore cheap, as broadly it was in the advanced industrial countries in the 1950s and 1960s, the incentive to monitor and limit its use was at best marginal. But since the 1970s, many governments have established agencies and/or formulated schemes designed to reduce demand by ensuring that energy is used more carefully, by means, for instance, of more efficient engines and boilers, and through better insulation. The potential for energy-saving in these and other ways is undoubtedly great, especially in those societies which have been particularly profligate in the past in their energy consumption, the United States and Canada above all. Conservation is also, in general, environmentally attractive and can now be very cost effective. But the difficulties in realizing its inherent potential are considerable. In the first place, whereas supply policies can be concentrated on a relatively few facilities and decision-makers, the construction of electricity-generating stations being a good example, conservation policies must ultimately reach out to large numbers of consumers. Energy prices are accepted as the foremost way of doing this, and public education campaigns are also favoured, but there is much still to be learned about the most effective procedures. Other uncertainties with conservation policies concern their absolute contribution to reducing energy demand and their true long-run cost effectiveness. Thus, for instance, given that one effect of better insulation may be to free finance for other energy-consuming activities, the initial energy benefit is obviously reduced by these new demands. Again, if one consequence of energy conservation is, as is conceivable in some circumstances, to discourage innovation, then this also needs to be allowed for in calculating the net gain. It would seem a fair conclusion that while the potential of conservation is large, achieving real, substantial and continuing savings is neither straightforward nor easy.

The British case

This chapter has thus far dealt with the problem of energy in general. It is desirable now to focus on the British situation, not to delineate it completely but rather to illustrate the complexity of decision-making in the energy field in just one country.

The discovery of North Sea oil and gas has left Britain well endowed, at least in the medium term, with energy resources, given the country's large coal reserves and nuclear power capability. The main facts, in outline, are as shown in Box 1.

The continuing importance of coal is clear, as is that of the remarkably rapidly established domestic oil and natural gas industries. It is also apparent that the role of nuclear power in Britain is still a limited one, despite the expenditure which there has been on it, and the public attention which it has received. Renewable sources evidently remain highly marginal, and this will remain true for the forseeable future.

But how is an energy policy to be constructed from these components? In the words of the Energy Policy Review of 1977:

The objective of energy policy is to secure that the nation's energy needs are met at the lowest cost in real resources, consistently with achieving adequate security and continuity of supply, and consistently with social, environmental and other policy objectives. The nation's future energy needs are not immutable. (*Energy Paper No. 22*, HMSO).

Britain's evident good fortune should not obscure the underlying issues. Strategy must aim at a robust flexibility — supply diversification combined with demand modification where the latter is clearly economically indicated. In turn, these broad policies call for inescapably political decisions and definite technological progress. Ironically, there is a danger that Britain's energy riches in the medium term will conflict with the steps needed to accommodate the longer term (in which domestic oil production will be declining).

Box 1

Britain's energy position

Self-sufficiency in energy production was achieved on a net basis in 1980 and this position will be maintained at least for most of the 1980s. Peak energy consumption to date occurred in 1973, the last year before the first oil crisis. Britain's energy consumption of primary fuels in 1980 was as shown in the table (1 tonne coal = 0.6 tonne oil = 250 therms)

	Thousand million therms	*Million tonnes coal equivalent*	*Million tonnes oil equivalent*
Coal	29.1	120.9	71.1
Petroleum	30.7	121.4	71.4
Natural Gas	17.6	70.4	41.4
Nuclear electricity	3.0	13.3	7.8
Hydro electricity	0.5	2.0	1.2
	80.9	328.0	192.9

(Based on *Digest of UK Energy Statistics 1981*, HMSO)

130.1 million tonnes of coal were produced in the year from some 213 mines, and power stations consumed 89.6 million tonnes: there were some 228,000 wage earners on colliery books, this figure being almost a third of the total employed in the energy industries. With a refinery capacity of 130.1 million tonnes, output was 79.2 and total inland deliveries of petroleum products 71.2, of which motor spirit comprised 19.1 and fuel oil 19.2, power stations taking 6.5 and all road transport 25.0. 44.78 million tonnes of crude petroleum were imported and 14.11 million tonnes of refined petroleum, exports being 38.46 and 16.05 respectively. There were 260 continental shelf oil and gas licenses in force, twenty-six oil fields were being developed and fifteen of these were in production. 79,113 million tonnes of oil were produced and 37,290 million cubic metres of gas, reserves in existing discoveries being 2,300 million tonnes and 1,559,000 million cubic metres respectively. Total gas sales were 16,630 million therms, of which domestic consumers took 8,439 million and of which town gas comprised only 31 million. 247.7 terawatt hours of electricity were supplied, 29.2 of them being of nuclear origin. Nuclear made up 5,767 of the total output capacity of 68,311MW and maximum demand was 49,467MW. In terms of fuel input to the electricity system, of a total of 114.7 MTCE coal and coke made up 89.7, oil 11.2, nuclear 11.9, hydro 1.7 and natural gas 0.2.

The objective of flexibility contrasts sharply with the comparable outlooks of, for example, the mid-1950s and the mid-1960s. In the mid-1950s, an energy crisis was thought imminent and all the eggs which could be were placed in the nuclear basket. The crisis did not materialize, or more accurately it was delayed by almost two decades. In the mid-1960s, the mood became one of astonishing confidence, of being able to plan in a predictable environment. The 1970s, having thoroughly destroyed these expectations (naive at the time and not only in retrospect), the commonsense approach came to seem the provision of alternatives to cover a wide spectrum of possible futures. Such provision naturally imposes its own costs both direct (for example, electricity-generating capacity well in excess of normal margins) and indirect (for instance, social irritations at initiatives which seem premature, above all a rapid nuclear build-up). It also creates its own momentum, pre-empting national resources for the energy sector, encouraging expectations in each of several energy industries, and in the end, despite the best intentions, imposing its own constraints. What cannot be too often said is that despite its own strong energy position, Britain cannot isolate itself from the wider, and far more turbulent, outside world. The country is far too dependent on foreign trade for that to be possible.

Energy decisions obviously have a significance which extends well beyond the energy sector itself. Thus, a state dependent on foreign sources, whether of oil or uranium, needs must reflect this dependence in its foreign policy. Or again, a state which has made itself dependent on nuclear power, unavoidably a highly centralized energy source, will have inevitably to allow for this administratively in a distinctly different way from one which has chosen, or been able, to rely on smaller scale technologies. A politically important subject is commonly given its own government department. Such, as noted above, has been the case with energy since 1973 in Britain and elsewhere. Earlier in Britain, the sector had been designated 'Power' or 'Fuel and Power' and the responsible minister had not always even been in the cabinet. Embodying a subject in a department usually allows policy effort to be concentrated, but it is well to remember that interdepartmental boundaries still exist. In the case of energy, the two most important boundaries are those with environmental and industrial questions (these two subjects having had their own departments for some years in Britain at the time of writing). The departmental philosophy will not necessarily or even usually be the same in these other departments as it is in Energy, and this can be the source of conflict. Having that part of manufacturing industry concerned with the construction of energy facilities inside the Energy Department, as having all the separate energy supply industries themselves under one official roof, removes interdepartmental problems but substitutes intra-departmental ones. Does one hold a public enquiry and accept delay, or push ahead with an energy development which seems urgently needed? Does one order a new power station to keep the construction industry in business even in the face of excess existing capacity? These are not hypothetical problems but real ones, both of which had in fact to be faced in Britain in the late 1970s.

Looking to the future

Historically, and quite generally, energy consumption has been correlated with the level of economic performance, and there are also traditional national attitudes to its use. Thus, the *per capita* consumption of energy in the USA and Canada has run at some two to four times that in Western Europe, while figures for the latter are 25 to 100 times higher than those, for example, in the developing African countries. In the case of particular countries, the rate of growth of energy consumption has tended to bear a roughly constant ratio to the rate of growth of gross national product over quite long periods — around 0.6 for the United Kingdom, for instance, in the decades after World War Two. But these ratios are not unchangeable and we should note that they became established during the years of energy abundance when little attention was paid to them, and still less to the details from which they were aggregated.

The evolution of sound energy policy is not assisted by the relatively short time horizon, not to say limited vision, characteristic of political decision making. Decisions to develop a new mine, expand a nuclear power programme, bring in a new energy source and so on, all take upwards of a decade to come to fruition and their consequences thereafter persist for decades. In liberal democracies, even if a particular party government stays in office this long, individual ministers virtually never do. Of course, the politician's focus on the near future is offset both by his more permanent officials and by his own qualities of statesmanship, but it may not be completely removed. And since politics is itself a kind of oil which keeps societies functioning, it is unhelpful to complain at its undesirable side effects. One should also reflect that politicians respond to pressures as they see them. Future generations and foreign populations have no vote and little voice, and if their case no longer goes wholly by default, neither is it usually centre stage. The politician also likes fixes, the more straightforward and spectacular the better, which is a further contributing factor in the fate of glamorous technologies like nuclear power as compared with more mundane ones, like some of those associated with conservation.

Adequate energy supplies are essential for industrial production and for personal comfort and mobility. If the pursuit of them degenerates into an international scramble, the resulting political strains are more likely than most other possible developments to precipitate new global conflict, with all the peril which that would entail in a nuclear age (see Chapter 11). On the other hand, the earth's fossil and nuclear resources are large, if unequally distributed; the potential of new technologies from coal-burning to fusion appears great; and the scope for modifying energy practices in the developed countries, without threat to the intrinsic quality of their life styles, seems considerable. It is entirely fitting that energy, because the most fundamental of all natural resources, should pose the technical, economic and political difficulties which it does. The seventies provided us with fair warning and we shall have only ourselves to blame if we fail to solve the energy problems which that decade identified.

Questions for discussion

1. What are the appropriate goals of energy policy?

2. What, in general, are the principal constraints on energy policy making?

3. How, in a liberal democracy (like Britain) should decisions on energy policy be made? What should be the roles of Parliament, the Government, the energy supply industries, pressure groups, and the public?

Further reading

The Energy Papers published by HMSO are an invaluable series. Over forty have now been published covering all aspects of the subject.

A number of Command Papers (also HMSO) have been published in recent years in this field. Of these, the sixth report of the Royal Commission on Environment Pollution, *Nuclear Power and the Environment* (Cmnd. 6618, 1976), remains especially useful.

On the US, Robert Stobaugh and Daniel Yergin, *Energy Future* (Ballantine Books, 1980, paper) is valuable.

Deciding about Energy Policy (Council for Science and Society, 314, St Andrew's Hill, London EC4V 5BY, 1979, paper).

Gerald Foley, *The Energey Question* (Penguin, 1976).

Britain's Energy Resources (COI 166, HMSO).

The reports of the House of Commons Select Committee on Energy (established 1979) are of great help.

10

COMING TO TERMS WITH
THE NEW GENETICS

Edward Yoxen

Very dramatic and important changes in how we think about ourselves are now occurring, catalysed by what I call here the 'new genetics', although many other areas of science and medicine are also involved. These developments principally concern the diagnosis and prevention of hereditary diseases, by means that will soon include the directed modification of human germ cells, or genetic engineering. These choices will be upon us faster than we realize, and before we have sorted out the moral, legal and psychological problems involved. Prenatal diagnosis, that is diagnosis in pregnancy of medical problems in the developing foetus, has been technically possible for the last fifteen years: it is now offered to thousands of prospective mothers in the UK. For some conditions it is routine. Since 1978, several hundred children have already been born after *in vitro* fertilization, so-called 'test tube babies', and as their numbers increase the technical skills in manipulating human germ cells are improved and knowledge of the intricacies of development deepened. Genetic engineering has already been attempted twice, rather clumsily many would say, once to try and correct an enzyme deficiency, once with a fatal genetic blood disorder. Despite the lack of knowledge, twenty or more research groups around the world are now planning very similar things, to introduce normal genes into target cells in the bodies of patients to replace missing or defective genes.

Cloning of higher organisms, such as mice and rabbits, that is the production of genetically identical progeny without sexual reproduction, is now possible. The erroneous claim has been made that human beings have been cloned, partly to stimulate debate about this prospect before it becomes reality. Thousands of pounds worth of frozen calf embryos are now sold every year for reimplantation in host cows, and this business is likely to produce knowledge applicable to human beings.

Technical background

I have already begun to use some technical terms, and we should be clear what they mean. We can think of genes as instructions. Development and growth arise through the orchestrated enactment of such instructions to build up a fully formed individual belonging to a particular species. Heredity is the process of transmitting a new set of instructions to constitute one or more members of the next generation. In sexual reproduction two partial sets of genes are recombined to produce one complete set. In asexual reproduction, used by some plants and now possible with other organisms with very special manipulative techniques, no shuffling of the genetic pack takes place. Genes, either individually or in linked, interdependent groups, specify the production of particular traits in organisms, some of which may be directly observable, like eye colour or skin pigmentation, some of which concern essential physiological processes occurring in specific cells, such as the breakdown of a chemical ingested in food or the production of a hormone necessary to regulate the functioning of the whole organism. Even the simplest organisms possess thousands of genes and it is worth saying that we have a very partial understanding of how genes act; that is to say, some systems seem very well understood, in others we can only speculate about what happens, while yet others remain to be discovered.

Nowadays we also know that genes have a material basis; they exist as physical entities. They are located on rod-like bodies (called chromosomes) in every cell nucleus, as structured sections of the substance, deoxyribonucleic acid (DNA). Virtually every cell in a muticellular organism like a human being possesses a complete set of all the genes for that organism. In any cell at a given moment in the life of the organism most of these genes will be silent or inactive. It is also possible that the gene for a particular trait will be missing or present in a defective or unusual form and it is this situation that lies behind the thousands of genetic diseases now known to us. Most of these are very rare, though the more common like cystic fibrosis affect one in every 2,500 live births (a rate of incidence which amounts to around 300–400 affected children per year in the UK). Some are only common among specific populations, like thalassaemia, or Tay–Sachs disease. All in all, genetic diseases, including chromosomal disorders such as Down's syndrome, appear in around two per cent of live births. This is a significant medical problem and means that the prospect of their occurrence affects a considerable proportion of people wishing to have children.

Within the last fifteen years it has become possible to perform tests during pregnancy that will indicate with almost complete accuracy whether the developing foetus will suffer from a particular condition. This antenatal diagnosis is done by analysing foetal cells obtained at present by extracting some of the fluid surrounding it in the mother's womb, and this procedure, known as amniocentesis, carries a slight risk. In the future it may be superseded by less invasive techniques that can be used earlier in pregnancy, when termination is easier to perform if the diagnosis is positive. Even then such a procedure would still be unacceptable to some people on moral grounds. Whatever one's views are about the morality of this technique, it would be wrong to ignore the facts that many people are very concerned to make sure that their child will not have a genetic disease, but that getting this reassurance prior to birth can be a psychologically and legally complex process.

It is still the case, however, that many people are born suffering from genetic diseases and scientists are working on the correction of these defects by supplying missing enzymes or genes. This is highly experimental medicine, and it is debatable whether it should be attempted even for those whose chances of survival are nil. An even more difficult kind of genetic engineering for which the initial experiments have also been attempted, involves the heritable correction of genetic defects, by introducing genetic material into germ cells so that the 'correction' is then passed on to subsequent generations. This kind of manipulation of very delicate, microscopic cellular material is certainly facilitated by work on *in vitro* fertilization, which requires the removal of an ovum from the prospective mother. Fertilization and the very early stages of development take place outside her body and the growing blastocyst is then reimplanted in her body, although in principle it could be transferred to a surrogate mother. Human cloning would also require such reimplantation. Clearly then, even though we have moved here from routine procedures, through experimental work to possibilities that lie in the near future, we are discussing very challenging and problematic procedures.

Almost always, genetic diseases manifest themselves because the affected individual acquires one or two flawed copies of a particular gene from his or her parents. With so-called recessive conditions two copies of the gene must be inherited, one from each parent, both of whom will be perfectly healthy. They are often referred to as 'carriers', can be identified as such for a growing number of conditions and can be advised how to deal with the risks of having affected children. Such advice falls under the heading of genetic counselling. Another way of describing this is how to say that such individuals are heterozygous for a particular condition. One can also say that on average 25 per cent of the children of a couple who are both heterozygous for a particular condition will inherit two copies of the gene causing that condition and will suffer from it, 50 per cent will be carriers, and 25 per cent will inherit no copies of the gene. For some diseases 'carriers' can also be identified on a mass scale, and this is known as genetic screening. If carriers know their status prior to marriage and the appearance of children, they can consider whether they wish to find another marital partner, who does not carry the relevant gene, or whether to have children and to arrange for antenatal diagnosis, if it is available, and terminate the pregnancy if the foetus is affected.

Value choices

Already it should be apparent that some of the choices which arise out of the application of new genetic knowledge to medicine must be very difficult to make in practice, and directly or indirectly challenge some of the conventions and values of our society. I want to illustrate here the values that are implicit in present practices, that lead to certain kinds of choices for people, and that are implicit in research that is being done, that will be done or that could be done but may not be.

Let us consider the example of the identification of carriers of recessive genetic diseases. Firstly, it is worth pointing out that a concern with such conditions only arises in specific social contexts, where deaths in all age groups from infectious disease have been brought to a low level by improvements in medicine, and where family size is small and infant mortality low, so that people's expectations of a healthy life for themselves and their children have risen. Secondly, genetic diseases are usually not immediately lethal: a long period of debilitating illness can be expected in many cases. If death were inevitable a few minutes after birth, it might be that concern with prevention would be less, at least in countries with a relatively high infant mortality. Correspondingly, then, identification of carriers takes place because of the judgement that it is worthwhile trying to alleviate the distress caused by the conception, birth, illness and death of affected children, through the identification of individuals whose children are likely to have a specific disease if the other parent is also a carrier. Note that if the other parent is not a carrier of the relevant recessive gene, then the only problem that arises is that on average half their children will be carriers, but generally this is not thought to be an important consideration. But even if carrier identification is useful to the individual, is it so important to the community, that screening of entire populations is indicated? If it is, should such genetic screening be obligatory? The relevant issues here are cost and priorities in health care, individual rights to privacy and individual obligations to some larger social group, and the moral implications of acting upon the knowledge gained. The prevailing view is that some genetic screening is important, but that there are problems in making it obligatory. Continuing research is likely to make it cheaper and applicable to a wider range of conditions. At the present time genetic screening for the blood disorder sickle-cell anaemia is compulsory in the majority of the states in North America for black school-children. There are voluntary community-based screening programmes for Tay—Sachs disease in the United States, but not in Britain, for Ashkenazi Jews. The question of screening of school-children for carriers of the cystic fibrosis gene has already been raised in anticipation of the as yet unrealized technical possibility. In a moment I shall consider the psychological and social problems that may be caused by such screening programmes; before that I want to consider the values that underlie the choices offered to carriers identified by such programmes.

Carriers may be adolescents in a sensitive phase in the formation of an adult self-identity. In this case they will have to come to terms with this threat to their self-image and decide how to choose and how to present themselves to sexual and/or marital partners. They may be married, intending marriage, or in a stable relationship intended to last, but without children; or they may have children who are unaffected, but who may themselves be carriers. I am considering here the situation for people in stable relationships. For single individuals the situation is a little different. I am assuming, moreover, that single parenthood is likely to be a state attained after the breakdown of a relationship, rather than one maintained intentionally throughout. In modern industrialized societies, even though family size has now fallen, the raising of children of whom marital or conjugal partners are the biological parents is highly valued, and has a major effect on people's image of themselves. People who find that they and their marital partner are both heterozygous for a serious genetic condition are therefore very likely to maintain their desire to have children. At some point they will have the following options:

(1) Remarriage to someone known not to be a carrier of that gene.

(2) Extra-marital intercourse, with someone known not to be a carrier, to conceive children who would be brought up by the married heterozygous couple.

(3) Artificial insemination by an unknown donor (AID) of known genetic status.

(4) Surrogate motherhood, with the return of the child after birth to the biological father and his wife.

(5) Conception in the 'conventional' way, followed by antenatal diagnosis and termination of pregnancy if the foetus is affected.

(6) Adoption.

(7) *In vitro* fertilization, followed by gene therapy to correct the genetic defect at a very early stage of development.

(8) Cloning of one member of the couple.

(9) Surrogate parenthood, following embryo transfer to the woman concerned so that the couple experience a pregnancy directly, even though neither of them would be biological parents of their child.

Let me consider each of these in turn. The first response is perhaps more likely for those contemplating marriage. For some people divorce violates the sanctity of marriage and requires the renunciation of commitments made in the wedding ceremony. Alternatively it could be the long-term solution, if other courses of action fail. It may become more acceptable psychologically as the divorce rate increases.

The second response would also be morally unacceptable to

some people, and might seem psychologically unacceptable to many more. However it is known from blood group studies that conception following extra-marital intercourse is much more frequent than one might suppose (estimates run as high as 30 per cent of all pregnancies), so that a not insignificant number of women must know that their spouses are not or may not be the biological fathers of their children, even though the men may not know. This data says nothing about the present psychological effects of this situation, which are unknown; nor is it directly relevant to the situation where conception by another partner is deliberately and overtly sought; where there is a 'medical' reason for such action, it might be easier to accept. What would be necessary for this to work is a revaluation of conventional attitudes to monogamy and sexual exclusiveness, a process that would take time, as it would require changes in a fundamental aspect of socialization.

The third possibility avoids some of these psychological complications although the evidence available suggests that people do not find AID easy to live with and its legal implications have not yet been fully worked out.

The fourth option is certainly possible now. Cases of surrogate motherhood are occasionally reported, particularly when they lead to court cases. A physician in Kentucky is now offering surrogate motherhood on a commercial basis.

The fifth option is the most frequent at present, but some people find abortion unacceptable on moral grounds. Even those people who opt for it are likely to find the period in which the diagnosis is established, which may last several weeks running up to and perhaps beyond the legal limit for the termination of pregnancy, very stressful emotionally. In practice, antenatal diagnosis is more complex technically than I have implied, as the accuracy may be less that 100 per cent and for some conditions with a different mode of inheritance from that assumed in my example (e.g., sex-linked recessive conditions without antenatal indications) the only course may be to abort all foetuses of one sex, of whom 50 per cent will be normal. One course of action that alleviates some of these problems is to do more research on antenatal diagnosis, so that it can occur earlier in pregnancy, and on gene therapy, but this will require research with foetal material, possibly obtained by *in vitro* fertilization, and to some people this is unacceptable morally. A moratorium on foetal research was indeed imposed in the State of Massachusetts in the mid-1970s. Another possibility is to develop post-natal gene therapy, or to improve the treatment for particular genetic conditions. But, on a different moral level, one has to justify the allocation of resources to deal with conditions that should be preventable.

The sixth option seems more straightforward, but here the problem is that the numbers of children, particularly infants, continue to fall. Indeed, there is already a black market in babies from less developed countries.

The remaining three possibilities are not yet feasible. They may seem outlandish, but I make no apologies for including them, because they will certainly become possible technically. If they are not taken up in practice — and I am not necessarily suggesting that they should be — then it will be because of decisions made in the courts, in parliament and by professional bodies such as the Royal Colleges of Medicine, to classify these practices as unacceptable, to withdraw resources from them, and to police a system of controls that prevent individuals from taking these options. I suggest that, unless the general desire to have children weakens and a major shift occurs in attitudes to women, then these possibilities, particularly surrogate motherhood, will be available to the wealthy and desperate. In my view it is more important to tackle the problem of why people have come to feel so desperate that they must have children; more important to ask what is likely to be the economic situation of a woman prepared to act as a surrogate mother on a commercial basis; more important to make available sensitive, non-repressive and supportive ante-natal and obstetrical care; and more important to place in women's hands a real control over fertility.

Difficulties with screening

These problems will be upon us very soon, given the pace of research. But at present the situation is that genetic screening occurs to detect heterozygotes among special sub-populations and, in pregnancies, to detect specific non-recessive defects in high risk groups, such as mothers over the age of thirty-eight who are routinely offered antenatal diagnosis to pick up cases of Down's syndrome and mothers whose children are known to have a higher than average likelihood of having the condition, spina bifida. There are problems associated with the heterozygote screening programmes, which I want to mention briefly.

The discovery that one carries the gene for a specific condition, which could affect one's children even though it has no effect on one's own health, can be traumatic because, as I have just demonstrated, it implies that different kinds of reproductive choice have to be taken. The path to parenthood will be different and some people feel that this reflects on their essential worth as a person. Their self-image is compromised and they feel stigmatized. It requires sensitive counselling to alleviate this problem. It can of course be made worse if heterozygosity is interpreted as a health defect in carriers. In the case of sickle-cell anaemia this has indeed occurred, because of an existing pattern of discrimination against black people, among whom the condition is much more common. Carriers have found themselves debarred from some kinds of employment and penalized by having to pay higher health insurance premiums in the United States. Only recently has the US Air Force agreed under pressure not to exclude carriers from pilot training, because of the alleged possibility that they would be more prone to respiratory failure in moments of oxygen shortage.

A screening programme to deal with sickle-cell anaemia in a rural Greek population has shown the kind of problems that stigmatization can cause. All individuals in a community of around two thousand people were screened; retrospective surveys showed that virtually all of them knew their own genetic status after the test and understood its medical significance. Yet in a follow-up study, to the surprise of the researchers, it was found that a significant number of people had been hurt by the breakdown of marriage negotiations into which this data had been introduced at the last moment. In other cases the genetic status had been concealed from the spouse and his or her family, with the result that the incidence of the gene in the community was the same as it would have been had no screening programme taken place. The lesson to be learnt here is that data about genetic status is never just that. It is also given a meaning and significance that is specific to the culture in which individuals live. They come to terms with this medical knowledge in ways that depend upon culturally transmitted ideas about health, marriage, parenthood, family honour and desirable attributes for men and women. These ideas vary markedly from one culture to another; and any counselling that ought to accompany screening has to be based on this kind of understanding of genetic disease.

In the case of all the conditions mentioned so far – cystic fibrosis, sickle-cell anaemia, thalassaemia – and a long list of others, the causal connection between being born with the necessary one or two copies of the gene and manifestation of the disease symptoms is straightforward. In any environment that trait will be expressed. In a few cases, very special manipulation of the environment, e.g., to remove foods containing the amino acid phenylalanine, can prevent the development of the disease. But in a growing number of cases the connection is more complex, particularly where the hereditary trait is a predisposition to behavioural disorders or to develop conditions like cancer. In these cases the model of causation is often extremely vague and several different courses of action could follow from its acceptance. Thus in the case of supposed hereditary hypersensitivity to toxic chemicals and carcinogens, it is clear that the supporting evidence is very poor and could not yet justify the establishment of a screening programme, for example to protect workers held to be particularly at risk from the chemicals used in a factory. Even if the data were more reliable, it is still not obvious that workers' health is best protected by this kind of selective exclusion from harmful workplaces.

Similarly work is now going on to try to identify hereditary factors behind mental disorders such as schizophrenia. Even if one were able to show that certain genes appear more frequently amongst schizophrenics and their families, it is still not clear that this would be very helpful given the multiplicity of other factors that influence the formation of personality. Indeed, to screen for such a gene, in the absence of any understanding of how it interacted with other influences, would seem to be decidedly unhelpful. In the next few decades, however, we can be certain that a great deal more knowledge about human genes will be discovered and we shall be forced to ask ourselves what significance shall these items of information have and how do we wish to act upon them.

Resolving the issues

Finally, it is worth considering how societies like that in Britain can come to terms with these new possibilities and dilemmas. Other societies, with different traditions of government, population densities, relations between church and state or levels of economic development, will respond in their own ways. Not many countries will be so poor and so isolated that they can totally ignore these problems.

Genetic screening, antenatal diagnosis and gene therapy are, or will be, medical procedures, handled by professional people with a strong sense of the need to defend their autonomy as professionals and to be bound as far as possible only by moral injunctions formulated by their professional peers.

Undoubtedly the specific moral problems of genetic engineering will be considered by such august bodies in due course. Also, medical practice, particularly of a controversial or experimental kind, is supposedly monitored by committees of doctors in hospitals, with the power of veto over particular experiments. Any medical procedure is also supposed to be sanctioned by informed consent of the patient and/or his or her representatives. In practice both these sets of controls are often abused, and they presuppose the existence of some kind of moral consensus and a general awareness of the problems involved. That level of awareness has to be created, by argument in the media, by discussion in educational institutions, by political debate, surrounding particular pieces of legislation and by legal disputes of all kinds. Certain kinds of experiment or procedure could be prohibited, as with vivisection. Certain constraints and limits could be negotiated and given the force of law, as is the case with abortion. But whatever form the resolution of these issues takes, it must begin soon. In the United States a Presidential Commission is considering a whole range of these problems. The British approach is regrettably much less deliberate and public and much more occasional and private.

Questions for discussion

1. Assuming that it might make medical sense to screen whole national populations, generation after generation, to detect and notify the carriers of deleterious recessive genes, at what age and in what form do you think such information should be given to the people concerned? In your view, who needs to know information about genetic status? Do you think that GPs are likely to be able to give adequate counselling? What kinds of abuses might occur?

2. Do you think that freedom to reproduce is an absolute and primary right or do you think that in certain cases individuals may legitimately be prevented from passing on their genes? What kinds of criteria do you believe to be relevant to this decision?

3. Do you see the issues discussed in this chapter as leading to greater individual freedom and choice or to greater coercion and a restricted range of more complex choices? Why do you think you see the direction that developments are taking this way?

Further reading

Anthony Smith, *The human pedigree: inheritance and the genetics of mankind* (George Allen and Unwin, 1975).

A highly readable, discursive treatment of a whole range of issues connected with human genetics, including some fascinating historical material.

Charles Birch, Paul Albrecht (eds.), *Genetics and the quality of life* (Pergamon Press, 1975).

A series of contributions organized by the study group of the World Council of Churches, some with an explicitly theological treatment.

Amitai Etzioni, *Genetic fix: the next technological revolution* (Macmillan, 1973; Harper Colophon paperback, 1975).

Life and death before birth (Council for Science and Society, 1978). (Available from 3/4 St Andrew's Hill, London EC4 5BY).

A brief review of the issues surrounding amniocentesis and selective abortion. Clear, informed, meticulous in its attempts to see the problems from every angle.

Z. Harsanyi, R. Hutton, *Genetic prophecy: Beyond the double helix* (Paladin paperback, 1982). A fascinating account of the way in which genetics could transform our views of many diseases. Informed opinion is that they exaggerate the present utility of a genetic perspective, but this is still a thought-provoking book.

11

SCIENCE, TECHNOLOGY AND THE BOMB

Philip Gummett

In some sort of crude sense, which no vulgarity, no humour, no overstatement can quite extinguish, the physicists have known sin, and this is a knowledge which they cannot lose.

Although not pronounced until some time later, these words of Robert Oppenheimer distilled the essence of the moral concern which assailed the nuclear weapons scientists in the United States in 1945, and which led them to a variety of distinct positions. After the defeat of Germany, some scientists decided that there could be no further need for the bomb and left the Manhattan Project, while others became concerned about whether there was any need to use it against the Japanese. In the end, however, and on the advice of a high-level committee which included three eminent scientists and had a distinguished scientific advisory panel, the President concluded that the alternative of a military demonstration, perhaps on an uninhabited island, might fail to end the war and that the implications of an invasion of Japan were even worse than the use of the bomb. At about the same time there arose the question of how to achieve international control of nuclear weapons, and whether this involved telling Russia about the bomb before it was used; not to do so might place a threat of duress over any subsequent attempts at control. This concern led some scientists to place their duty to the international order above that to their own country, and hence to pass information illicitly to Russia.

Moral dilemmas remained after the war. In Britain there was the question of whether to assist with the British bomb programme. In the United States, in what eventually led to the Oppenheimer hearings, there arose the debate over whether to try to build the hydrogen, or fusion, bomb — a weapon that would be immensely more powerful than the existing atomic, or fission, bombs. As Margaret Gowing (1979) summarizes this debate, two questions arose over the H-bomb: first, whether it was scientifically and technically possible and secondly, if the first answer was yes, whether the USA should for political and moral reasons go ahead. One group of scientists, she continues, felt passionately that if enough resources were poured into the project a hydrogen bomb could be made, and that in view of the Communist threat it must be made. The Atomic Energy Commission's scientific advisory committee (chaired by Oppenheimer) opposed this view. They believed that the weapon probably could be produced, but at the costs of diverting effort from the development of better fission bombs and of producing an intolerable threat to the future of the human race. They were sure that American superiority in fission weapons was so great that renunciation of the H-bomb, in the hope that Russia would follow suit, would not endanger American security.

H-bombs were developed, and the prospect now exists that any country attacked by them will suffer unimaginable devastation by blast, fire, immediate thermal and ionizing radiation and, in the aftermath of the explosions, radioactive fallout (this last, of course, carrying the potential of affecting countries far from the centres of attack). Space does not allow a discussion here of the effects of nuclear weapons, but the Appendix summarizes key data on this appalling subject in the belief that without an appreciation of these effects, any discussion of the nuclear arms race must be incomplete.

By no later than the mid-1950s, however, the focus of the nuclear arms race shifted from the development of 'better' bombs (although refinements, such as the 'neutron bomb', have continued to be made) to the means of delivering them to their targets. The key questions in nuclear strategy became the numbers and types of 'delivery vehicles' and the credibility of the threat — essential to deterrence theory — that, regard-less of anything the adversary could do, sufficient of these vehicles would get through to cause unacceptable damage to the people and industry of that country. The initial emphasis was on bombers, then on land-based and, later, submarine-launched ballistic missiles, with the long-range cruise missile being most recently added to the armoury.

These developments in the technology of weapons-delivery have created enormous problems for the control of the arms race, to which the next part of this chapter is devoted. But it would be wrong to suggest that technology alone is the 'cause' of the arms race — as if anything so complex could have a single cause — and so we will also see briefly how a fuller understanding of this question might locate technology alongside other key factors. We begin by taking as an example the most serious step to date towards nuclear arms control, the Strategic Arms Limitation Talks (SALT).

The Strategic Arms Limitation Talks

The SALT talks began in 1969. One reason for them was that both sides were concerned about the steeply rising cost of de-fence and the prospect of new, and destabilizing, weapon sys-tems. There was especial concern about plans for antiballistic missiles (ABMs) — missiles designed to attack incoming ballistic missiles high in the atmosphere — and multiple, independently targetable re-entry vehicles (MIRVs) — that is, the capacity to put onto one missile between, say, three and fourteen warheads each of which could be aimed at a different target. ABMs were seen as destabilizing because, if one side could defend itself against a missile attack, deterrence might break down. And the problem with MIRVs was that they rapidly increased the number of deliverable bombs, allowing multiple attacks on key targets and thus greatly increasing the chance of destroying them even if they were enemy missiles in strengthened concrete silos; thus MIRVs opened the way to fears of a pre-emptive 'first strike' against the enemy's missiles.

There were further reasons for SALT that were specific to each side. The United States was anxious to freeze Soviet deployment of strategic weapons, which were threatening to leave the United States behind. The Soviet Union was anxious to gain international recognition that it had 'arrived' at super-power status, equal to the United States at least in military (if not economic and technological) terms, and participation in these exclusive talks was a symbol of that arrival.

What was achieved in the first stage of SALT, known as SALT-I? The treaties signed in 1972 did two things. (See Box 2). First, they banned the nationwide deployment of ABMs in the USA and USSR, limiting to two each the number of ABM sites that would be permitted. Second, they banned for the duration of the treaty (five years) the deployment of any new intercontinental ballistic missiles (ICBMs) or submarine-launched ballistic missiles (SLBMs), except those already under construction and except also under the terms of a provision whereby new SLBMs could be deployed in exchange for dismantling ICBMs.

The psychological and political importance of the agree-ment on ABMs was considerable, for it meant that neither side was going to seek massive defences against missile attack, and therefore neither side was going to challenge, in this respect at least, the underlying stability of deterrence. But it must be observed that there was great scepticism among

---

Sources

International Institute for Strategic Studies (London), annual publi-cations, *The Military Balance* and *Strategic Survey*.

Gregory Treverton, *Nuclear Weapons in Europe* (IISS, Adelphi Papers No. 168, 1981), Appendix.

Statement on the Defence Estimates 1980 (HMSO, Cmnd. 7826-I, 1980), Fig. 2.

Statement on the Defence Estimates 1981 (HMSO, Cmnd. 8212-I, 1981), Fig. 2.

Note: It is possible to find variations on the figures given here for strategic forces, depending on quite what is counted. The figures given here are a good cross-section of those available.

	Box 2	
USA and USSR strategic forces and SALT		

US & USSR Strategic Forces, 1970 & 1980		SALT-I Limits (1972–7)		SALT-II Limits from 1 January 1981 to 31 December 1985	
1970	*1980*				
ng range bombers		*Bombers*		*Total strategic delivery vehicles*	2250
400	338+	Not covered by treaty		of which	
SSR 140	156			*MIRVed*	1320
bmarines		*Submarines*		i.e., ICBMs, SLBMs and bombers firing cruise missiles)	
41	41	US	44		
SSR 14	71	USSR	62	of which	
BMs*		*SLBMs*$^{++}$		*MIRVed missiles*	1200
656	656 (496)	US	656–710	of which	
SSR 304	950 (192)	USSR	740–950	*MIRVed ICBMs*	820
BMs*		*ICBMs*			
1054	1052 (550)	US	1054	Subsidiary agreement (expired 31 December 1981)	
SSR 1513	1398 (750)	USSR	1618	No deployment of ground or sea-launched cruise missiles with ranges over 600 km.	
tal delivery vehicles* ombers, SLBMs & ICBMs)		*ABMs* 2 sites each, 100 missiles per site (reduced in 1974 to 1 site each, and US has not even fully taken up that option).			
2110	2046+ (1046)			no tests of MIRVed cruise missiles with ranges over 600 km.	
SSR 1957	2504 (942)				
tal warheads and bombs					
4000	9200			Backfire bomber (USSR) not to be used at intercontinental ranges.	
SSR 1800	5500				
tal throw weight illion lbs.)					
–	7.2				
SSR –	11.3				

Numbers MIRVed in 1980 in brackets
Excludes some 200 B-52's in storage (but which are counted towards the SALT-II total of 2250)
+ Higher SLBM numbers allowed if difference met by dismantling older ICBMs.

experts about the feasibility of building a worthwhile ABM screen. This scepticism was based party on the delicacy and complexity of the electronic control systems that would have to respond largely automatically to radar sightings of ICBMs, programme ABMs on to targets, launch them, and all this in time for the ABMs to gain sufficient height not to damage their own country when they fired their own nuclear warheads. Suppose the radar was faulty, or the computer software failed, or the enemy successfully created a radio blackout? Additional doubts were raised by the observation that it would always be easier and cheaper to increase offensive forces than defensive ones, therefore the ABMs would always face being overwhelmed; and partial success would not be good enough when it came to stopping nuclear missiles. These doubts presumably eased the negotiators' task, but it should not be thought that they rendered the treaty trivial. Certainly in the USA, and probably also in the USSR, strong lobbies in favour of ABMs existed, and it is likely that without the treaty significant ABM forces would have been deployed.

The other aspect of SALT-I, the freeze on ICBMs and SLBMs, also needs comment. In the period immediately before, and during, the negotiations, the Soviet Union was busily constructing ICBMs and SLBMs; the United States was not, and therefore achieved what, as has been said, was one of its chief objectives in the talks, namely a freeze on further Soviet construction beyond what was already in production. The terms of the treaty allowed the USSR many more ICBMs and somewhat more SLBMs than the USA. But these figures alone conceal the facts that whereas the USSR had the capacity, with its missiles, to deliver a greater *weight* of bombs than the USA, the USA, throughout the five years of the treaty, could expect to rise to about three times the Soviet *number* of warheads, an increasing number of them independently targetable (MIRVed).

This observation leads on to what was widely seen as one of the greatest shortcomings of SALT-I, and to which the subsequent effort known as SALT-II was devoted. Under

SALT-I, no limitation was set on qualitative improvements (as opposed to sheer numbers) in missiles. Thus, under the 'ceilings' agreed in 1972, technological advance has enabled each side to increase the destructive capacity of its missiles. This has been achieved partly by MIRVing (about half the USA's missiles — counting ICBMs and SLBMs together — are now MIRVed, and about a third of the USSR's), and partly by increases in missile accuracy. In 1970 the CEP (circular error probable: the radius around a target within which fifty per cent of warheads aimed at it would be expected to fall) of the best US ICBM was about 650 m; this had fallen by 1980 to 200 m and the proposed MX missile is likely to have a CEP of better than 80 m. Since the 'lethality', or 'kill-probability', of a warhead varies as the inverse square of CEP (so that halving CEP raises the lethality fourfold: $1/(\frac{1}{2})^2$), a modest increase in accuracy considerably increases the chance of destroying, say, a missile silo with one shot.

One of the main intentions behind the SALT-II negotiations was to try to deal with these advances in technology. The task was complicated by the arrival of yet more new weapons, particularly the US cruise missile and the Soviet 'Backfire' bomber. The cruise missile is a low-flying, airbreathing machine — effectively a small pilotless jet aeroplane — which is difficult to detect and which, because of a guidance system which allows it to check its position against the contours of the ground below, is highly accurate (with a CEP of less than 30 m). It can carry nuclear or conventional warheads, and trade-offs between warhead size and fuel load allow variations in range. The Backfire bomber is an advanced supersonic aircraft primarily intended as a medium range bomber but able to reach the USA from the USSR if refuelled in flight. With both weapons the problem was, were they to be counted as 'strategic' weapons? The USA claimed that the Backfire bomber was strategic because of its potential range, but the Soviet Union insisted that it was only a theatre weapon, for use in Europe or China. A similar kind of argument occurred over the cruise missile. In both cases, there is no simple way (e.g., satellite surveillance) of checking on the purposes for which the weapons are deployed; nor is there any simple way of checking exactly how many missiles have been MIRVed or exactly what types of missiles (and therefore payloads and accuracies) are in what silos.

Technology, therefore, greatly complicated the SALT-II negotiations. The resulting treaty was signed in June 1979 and was to last until the end of 1985. It tried to deal with one of the technological developments, MIRVing, by setting limits on delivery systems in a sequence of 'chinese boxes': so many systems in total, of which so many could be MIRVed, and so on. (See Box 2). Overall, the Soviet Union would have had to dismantle about 250 old missiles to conform to the treaty, but both sides would have retained the freedom to add more MIRVs up to certain limits. No satisfactory solution was, however, found to the problem of cruise missiles and Backfire bombers, and these were dealt with under a secondary agreement, or 'protocol'. Under the protocol, long range ground and sea (but not air) launched cruise missiles were banned, and an assurance was given that the Backfire bomber would not be used intercontinentally. The protocol, however, was to expire at the end of 1981.

The subsequent history of SALT, namely, the opposition which it met in the US Senate, its withdrawal by President Carter after the Soviet invasion of Afghanistan, the criticism which it received from the new President, Mr. Reagan, and his attempts in 1981–2 to begin, not strategic arms *limitation* talks, but strategic arms *reduction* talks (START), need not concern us here. It will be more instructive to leave this specific example and consider what factors *other* than technology influence the process of arms control.

Further problems of arms control

It is easy to be impatient at the lack of progress in arms control without fully appreciating the complexity of the task. We have just seen how technological developments such as MIRV, missile accuracy, cruise and the Backfire bomber greatly complicated the SALT negotiations. We have also noted that shortcomings in another field of technology, that of satellite surveillance, place limits on the verifiability of any treaty governing those same developments. As if that were not enough, two further types of problem complicate arms control. These we may call political problems and problems arising from asymmetries between the two sides.

The political problems are of three types. First, there are electoral politics, to do with the ultimate problem for any democratic government of getting, and staying, elected to office. Arms control negotiations are often affected by elections with which they happen to coincide: we have seen, for instance, the political debates in the Netherlands, West Germany and Britain over the NATO cruise missile programme announced in 1979 affecting the course of negotiations on that subject.

Secondly, there are bureaucratic politics. These stem from the institutionalized conflict which naturally arises within a government when different elements of the bureaucracy, set up to oversee and safeguard different interests, find themselves dealing with different aspects of the same problem. Thus, a defence ministry, concerned with maintaining national security, is likely always to be suspicious of arms control arguments. The disagreement within the United States government in August 1981 over the decision to build the neutron bomb provides another example: the Pentagon was concerned to provide the best weapons for its troops; the State Department was more concerned about the effect of the decision on NATO allies in Europe who were already facing political pressure over the cruise missile programme. Also significant in this context may be the relations which develop over time between industrial interests and sections of the bureaucracy.

Thirdly, as the neutron bomb example suggests, there are alliance politics. Another example of these arises in the debate over the NATO cruise and Pershing missiles. Whatever the military need for these weapons, it seems clear that a major reason for the decision was to allay concern in Europe (especially West Germany) over the value of US security guarantees. In short, could the US be relied on to support western Europe, right up to the level of nuclear war if need be? The deployment of intermediate range American missiles in Europe was seen politically as a way of reassuring the allies: should such weapons ever be used, Russian territory would have been attacked by American missiles, making it hard for the war to remain confined to Europe.

The other type of problem which affects arms control arises from asymmetries between the positions of the negotiating parties. These may firstly be asymmetries in force structure, as when, in the negotiations over theatre nuclear forces in Europe that began in late 1981, the USSR had over two hundred SS20s in place, but no NATO cruise or Pershing missiles were due to be deployed before the end of 1983. Another example arose in SALT over the fact that each side deploys different percentages of warheads on different types of delivery system.

A second asymmetry is of geography, with one superpower *relatively* securely bounded by oceans while the other is surrounded by, as it sees it, hostile powers and also lacks as convenient access to the oceans as its adversary.

A more subtle asymmetry, and one to which I cannot do justice here, is of strategic doctrine. Traditionally, though this is changing, the USA has had a philosophy of 'deterrence by punishment'. Under the terms of 'mutual assured destruction', deterrence works by threatening an adversary with massive retaliation even if only a small part of one's own forces survive. Under this approach, there is no need to think much about the use of nuclear weapons in various kinds of less than all-out warfare. In contrast, the Soviet Union thinks in terms of 'deterrence by denial'. Without expecting nuclear war to be anything less than horrifying, the Soviet Union has nevertheless determined not to sit back and endure whatever is thrown at it, but instead to seek to minimize what, it accepts, will still be massive damage, by means of civil defence, the capacity to fight all kinds of war at all conceivable levels, and the capacity to launch first-strike attacks on US missiles, thereby limiting the number of them available for retaliation. It is easy to imagine how such fundamentally different interpretations of the word 'deterrence' can lead to difficulties at the negotiating table.

Further dilemmas   Technology, then, is only one of the variables which complicates the task of curbing the nuclear arms race. It has to be seen in the context of the degree of public support which politicians can win for various proposals, the workings of the bureaucracy, (and its relations with industry), alliance relationships, the lack of comparability between the forces of adversaries, and asymmetries of geography and doctrine. Nor is there any simple mechanical relationship between these variables: their interaction is organic and extremely hard to analyse.

Nevertheless, technology remains a key factor in the nuclear arms race, and hence dilemmas of the sort which troubled Oppenheimer and his contemporaries continue to concern scientists and engineers. For the defence scientist there are the questions of whether one should be willing to work on *any* weapon and whether, if one thinks that defence policy has entered dangerous channels, it is better to stay 'inside' government and seek to influence matters quietly, or to move 'outside', where one would be freer to act, but less influential. This dilemma can be particularly acute for senior scientific advisers. For non-defence scientists, and for ordinary citizens with backgrounds in science or engineering who feel that their specialist knowledge gives them particular insight into the technological aspects of the arms race, there are other dilemmas, but these are not significantly different from those which confront any citizen who is concerned about this subject.

In particular, there is the choice (exaggerated for the sake of exposition) between working, slowly and patiently, for multilateral progress in arms control, and going for more dramatic measures. Those who feel that arms control has rarely achieved anything significant, and has merely managed the arms competition without altering its dynamics, may be drawn towards the arguments for unilateral disarmament. This position is often augmented by the view that it is immoral to try to ensure one's own security by threatening mass slaughter.

Among the nuclear weapons states, it is only in Britain that significant support exists for unilateral disarmament, that is, the renunciation of Britain's own nuclear weapons. This position is often associated with calls for the removal of all NATO nuclear bases in Britain and also for European Nuclear Disarmament through a process which E.P. Thompson has called 'multilateral unilateralism' – unilateral steps by one side in the hope of evoking a response from the other, and so on. The slogan, 'unilateralists are multilateralists who mean it', expresses well the underlying argument that multilateral arms control is seen as a cover for the absence of real intention to control the arms race; unilateralists are prepared to take risks to demonstrate their seriousness.

But exactly what risks they are prepared to take is a matter on which opinions vary. The structure of a defence policy without the bomb has been the subject of an enquiry by the Alternative Defence Commission (based in the School of Peace Studies at Bradford University). The options they examined include: reliance on conventional forces; reliance on a system of territorial defence (through the training of a non-professional militia); and non-violent civilian resistance (organized non-co-operation with the invader, such as strikes). To these might be added the policy of offering no defence, in the belief that attack is unlikely, that the moral price of defence is too high, and that the USSR will not last for ever. Obviously, all these options carry risks (but so, note their supporters, does a policy of nuclear deterrence), and are politically highly sensitive. For example, the first might involve a higher defence budget and conscription, while the second would involve compulsory military training (and re-training) for all citizens.

All carry the risk of nuclear blackmail, but their supporters would argue that this is a blunt instrument of limited value, especially when world opinion rejects the idea of nuclear attacks on non-nuclear states.

In addition to the argument that these positions, risky though they are, are preferable to the risks of nuclear war, the other important argument underlying them is the ethical one: that it is immoral to threaten the use of nuclear weapons. Supporters of deterrence, and of multilateral arms control, reply thus to this argument. First, although no one can contemplate with equanimity the use of nuclear weapons, it must be remembered that their point is to forestall moves by the adversary which might lead to war. (As in chess, one seeks to prevent the development of threats rather than have to react to them when they arise.) And the prevention of war, given the horror that even a conventional war entails today, is surely a moral act − an argument which carries weight among defence scientists. Second, the sophistication of modern weapons means that it is possible to have a range of targeting options which includes relatively isolated military sites as well as major cities. Of course, attacks on military targets would inevitably kill many civilians, but it should not be thought that nuclear targeting entails only full-scale slaughter of citizens, with no intermediate options. Third, the horror of invasion by an alien power is often underestimated. Finally, of those who seek only European (or simply British) nuclear disarmament, it is asked, wherein lies the morality of abandoning one's own nuclear weapons while continuing to shelter under the American nuclear umbrella?

There is, then, no monopoly of morality in this debate: a variety of positions may be honourably maintained. The nuclear physics of the 1930s, transformed by the engineering feat of the Manhattan Project into workable technology, refined ever since in terms of warheads and delivery systems, and interacting constantly with political, economic and other forces, has brought us to the current pass. Politically and morally it is a position which few would have chosen as the starting point for the efforts which must continue to be made to prevent the nuclear destruction of civilization. But history does not present us with the blank sheet of infinite possibility; we start from where we are, and we must pursue, nationally and internationally, the political debate about which of the paths ahead we shall follow.

Questions for discussion

1.   Use the data in the Appendix to estimate the effects of a nuclear attack on your town or city. Remember to consider the likely arrival of rescue and hospital services. Consider also where your water supply comes from, what would happen to sewage, where food might be obtained, and what communications might be like.

2.   How do you regard the possible alternatives to nuclear deterrence for the foreseeable future? How do you, as a citizen, feel personally about your own role in any of the alternatives?

3.   Accepting the moral ambivalence of nuclear deterrence, how do you compare its morality with that of the alternatives?

4.   Is arms control a waste of time? If, on the contrary, you think that despite its shortcomings it is worth pursuing, how could more progress be made? How might you negotiate reductions in the 1980 arsenals of the USA and USSR? (See Box 2).

5.   Should scientists and engineers be prepared to work on defence projects? All defence projects?

Further reading

Margaret Gowing and Lorna Arnold, *The Atomic Bomb* (Butterworths, for Science in a Social Context, 1979).

Summarizes developments until 1953, with special reference to the origins and development of atomic weapons and their implications for the postwar world. Useful on the role of scientists in these matters. Contains an extensive bibliography.

Robert Neild, *How to Make up your Mind about the Bomb* (Andre Deutsch, 1981).

A good general sourcebook, by a former director of the Stockholm International Peace Research Institute, on the history and mechanics of the arms race, the effects of nuclear attack and British policy. Critical of much that the author describes, but written in balanced fashion. A very good introduction to the field.

Scientific American, *Progress in Arms Control?* (San Francisco: W.H. Freeman and Co. − orders *via* 20 Beaumont Street, Oxford, OX1 2NQ, England; 1979).

A collection of articles from *Scientific American.* A good primer on nuclear weaponry (including cruise missiles, the neutron bomb and MIRVs) and on arms control and non-proliferation policy. Contains many useful charts and photographs.

Lasrence Freedman, *Britain and Nuclear Weapons* (Macmillan, 1981).

Excellent political history of British thinking about nuclear weapons. Some wider reference to strategic problems and to the European nuclear debate.

Duncan Campbell, *War Plan UK* (Burnett Books, 1982).

A detailed review and critique of civil defence plans.

Defence Without the Bomb, Report of the Alternative Defence Commission (Taylor and Francis, 1983).

A serious and detailed analysis of non-nuclear options for the defence of Britain and Western Europe.

APPENDIX

Effects of a 1 megaton (Mt) nuclear explosion

Physical effects

(a) Blast

Does at least half the damage in a conventional nuclear bomb. Produces sudden rise in air pressure (measured as *overpressure*) and generates high winds. Detailed effects are a function of altitude of explosion. (Overpressure is conventionally measured in pounds per square inch (psi), and I abide by that convention here. To convert to SI units, multiply psi by 6.895×10^3 to give a result in pascal.)

Peak over pressure (psi)	Peak wind speed (mph km/h)		Typical blast effects	Radius of effect			
				Airburst (8000 ft 2.4 km)		Groundburst	
				(miles km)		(miles km)	
20	470	750	Reinforced concrete structures are levelled.	0.8	1.3	1.3	2.1
10	290	460	Most factories and commercial buildings are collapsed. Small wood-frame and brick houses destroyed and distributed as debris.	3.0	4.8	1.9	3.0
5	160	260	Lightly constructed commercial buildings destroyed; heavier construction is severely damaged.	4.4	7.0	2.8	4.5
3	95	150	Walls of typical steelframe buildings blown away; severe damage to houses; winds sufficient to kill people in the open.	5.9	9.5	3.8	6.1
1	35	55	Damage to structures; people endangered by flying glass and debris.	11.6	18.6	7.3	11.7

(*b*) *Fire*
(*i*) Maximum slant range (i.e., length of line from bomb to point on ground) at which ignition from an airburst occurs in clear atmosphere:
7 miles; 11 km.
(*ii*) Maximum radius of fireball:
airburst − 0.65 miles; 1.04 km
groundburst − 0.87 miles; 1.39 km

(*c*) *Initial nuclear radiation*
Approximate maximum slant range (metres) for various doses (airburst).

Dose (rads)	Gamma radiation	Neutron radiation
50	3,100	2,400
450	2,600	2,100
1000	2,400	1,900

(Note that dose varies sharply with distance: hence difficulty of accurate calculation of effects.)

(*d*) *Idealized fallout plume*

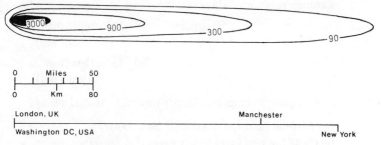

Figure 5   Main fall-out pattern

Note: The rem is a measure of biological damage; the rad is a measure of radiation energy absorbed. For our purposes they may be taken as equivalent.

Biological effects

(a) Blast

Blast kills directly, through certain effects upon the human body, and indirectly, through damage caused by the destruction of buildings or the effects of being caught up in hurricane force winds. In calculating blast fatalities, a rule of thumb is that survivors *inside* the five psi contour roughly equal fatalities *outside*. Hence, the approximate number of direct and indirect blast fatalities is equal to the population inside the five psi contour.

The data below refer to direct biological effects of blast.

Effect		*Occurs at average over- pressure of (psi)*	
Lung damage:	threshold	12	(8–15)
	severe	25	(20–30)
Lethality:	threshold	40	(30–50)
	50 per cent	62	(50–75)
	100 per cent	92	(75–115)
Eardrum rupture:	threshold	5	
	50 per cent	15–20	(more than 20 years old)
		30–35	(less than 20 years old)

Note: figures in brackets indicate variability around average.

(b) Thermal radiation (assumes 1 Mt airburst)
(i) Flashblindness (may be permanent) at ranges up to:
   13 miles; 21 km on clear day
   53 miles; 85 km on clear night
(ii) Skin burns at ranges up to:

1st degree	(similar to sunburn, with no blistering)	7 miles; 11 km

2nd degree	(blistering; can be serious)	6 miles; 10 km
3rd degree	(full skin thickness burn; treatment requires skin graft to prevent scar formation)	5 miles; 8 km

Note: 3rd degree burns over 24 per cent of the body, or 2nd degree burns over 30 per cent are usually fatal unless prompt, specialized care is available.

(*c*) *Effects of acute nuclear radiation* (i.e., doses received in relatively short time, say, up to one week).

May be received from immediate radiation from the bomb or from fallout.

Dose (rem)	*Effect*
600+	Almost certainly fatal within a few weeks.
450	Fatal for half the population exposed to it; the other half would get very sick but would recover.
300	Fatal for about 10 per cent.
200	Severe illness (diarrhoea, nausea, internal haemorrhage) but good prospects of recovery unless afflicted by other disease or infection while in susceptible state.
50	No noticeable short term effects, but will do long term damage in a small percentage (up to 2½ per cent) of those exposed.

Sources

Physical effects

(*a*) Office of Technology Assessment, Congress of the United States. *The Effects of Nuclear War* (Allanheld, Osmun; Croom Helm, 1980) — hereafter 'OTA' — p. 18, for airburst data; and Nuclear Bomb Effects Computer (available as supplement to Glasstone, *infra*), for groundburst data.

(*b*) S. Glasstone and P.J. Dolan, *The Effects of Nuclear Weapons* (United States Departments of Defense and of Energy, 1977) – hereafter 'Glasstone' – pp. 289–91; and Nuclear Bomb Effects Computer.

(*c*) Glasstone, pp. 334, 347.

(*d*) Based on OTA, p. 24.

Biological effects

(*a*) Glasstone, p. 552.

(*b*) OTA, p. 21.

(*c*) OTA, pp. 19–20.